Edible crab, plumose anemone, boring sponge, two small hermit crabs (one with its shell enveloped in orange sponge) and painted goby; all in a small patch of Devon seabed

GREAT BRITISH MARINE ANIMALS

3rd EDITION

PAUL NAYLOR

This book is dedicated to Teresa, whose love for the sea is so inspiring, and to Sam and Felix – a new generation of enthusiasts

Acknowledgements

Special thanks go to Cat Andrews, Ann Beeby, Lisa Chilton, Seb Shimeld and Mary Simmons for their valuable time, effort and expertise commenting on the new text and, also, to Tom Alderson, Ron Bird, Jolyon Chesworth, Ali Hood, Clare Howard, Anthea Preston and Nigel Smallbones for their assistance with earlier editions. I have greatly appreciated expert advice from Jack Sewell (and MBA colleagues), Claire Goodwin, Doug Herdson, Keith Hiscock, John Baxter and Rich Hurst on identifying animals in some of the photographs and am indebted to Alan Hodgson and Hazel Gooch for providing drawings and to Steve Carpenter at Sound Diving for his continued support. The help of many friends and fellow enthusiasts has enabled me to take the photographs; they include John Baldry, Bill Bowen, Mark Buddles, Anne-Marie Coriat, Cilla Course, Peter Glanvill, Terry Griffiths, Chris Holden, Les Kemp, Gary McClelland, John Newman, Tim Nicholson, Helen Nott, Gavin Parsons, Paul Parsons, Mark Perrott, Sue Scott, Sally Sharrock, Lyndon Taylor, Martha Tressler, Steve Trewhella, Robert Walker, Discovery Divers, Undersea Adventures, Plymouth Diving Centre and Porthkerris Divers (where special thanks go to Mike Anselmi for his patience and skill). Last but not least, thank you to Emily and Ellie for your encouragement, to Sam for your amazing enthusiasm in all weathers (and for finding a Montagu's sea-snail!) and to Teresa for all your expertise and wonderful support throughout.

The author

Paul Naylor has been snorkelling and diving around the coast of Britain for 35 years. Whether in the sandy shallows of a beach lagoon in Norfolk or in the deep and clear water around the Orkneys, he has always been enthralled by the animals that live there. For the last 25 years, Paul has concentrated on photographing these animals in their natural habitat and, wherever possible, showing them getting on with their lives despite his attention. His photographs have been widely used in campaigns for better protection and awareness of our marine life. Paul gives talks to universities, schools, government organisations, conservation groups and diving clubs on the wonders of our marine fauna; and writes about marine life for a variety of publications. He has a doctorate in marine biology and is an Associate of the Royal Photographic Society. This is his fifth book.

First Edition 2003
Second Edition 2005
Third Edition 2011

ISBN 978-0-9522831-6-4

Sales or other enquiries, including requests for photographs:

telephone 07041 351307 (local rate)

or visit **www.marinephoto.co.uk**
this site will also have information updates

Printed by Deltor in Cornwall:
www.deltoruk.com

CONTENTS

	FOREWORD	4
1.	INTRODUCTION	6
2.	SPONGES	25
3.	CNIDARIANS Sea anemones, corals, hydroids & jellyfish	36
4.	CRUSTACEANS Crabs, lobsters, prawns, shrimps & barnacles	84
5.	WORMS	124
6.	MOLLUSCS Chitons, sea snails, sea slugs, bivalves, cuttlefish, octopus & squid	140
7.	BRYOZOANS	190
8.	ECHINODERMS Starfish, brittle stars, sea urchins, sea cucumbers & feather stars	194
9.	SEA SQUIRTS	229
10.	FISH	238
	Glossary	308
	References for more information	310
	Organisations promoting conservation and awareness of marine animals	312
	Examples of different aspects of behaviour	314
	Index of species, English and scientific names	317

Foreword photographs: sea orange, sunset cup-corals, long-clawed squat lobsters, organ pipe worms, painted top-shell, purple sunstar, light-bulb sea squirts, short-snouted seahorse, Loch Carron (background)

FOREWORD

Our island nation is surrounded by seas teeming with life of great diversity, colour and intrigue. UK seas are so productive that well over 7,000 species jostle for space in every nook and cranny on the seabed around us, and within each cubic centimetre of the water column. Paul Naylor's book is a journey of discovery to uncover the world beneath the waves, its inhabitants and their remarkably colourful and often dramatic lifestyles.

As the UK charity whose mission it is to protect our seas, shores and wildlife, the Marine Conservation Society campaigns to protect sea life and habitats from the threats of over-fishing, pollution and neglect. Paul Naylor is a real champion of the cause - this book reflects the sheer wonder of the life on our shores and in our seas – life that we must all take action to protect, if future generations are to continue to be amazed by the animals so beautifully described in this guide.

Capturing the beauty, diversity and behaviours of hundreds of creatures for this book has been a phenomenal achievement. With its stunning photographs and engaging narrative, this is so much more than an ID guide - it is an inspiration to armchair naturalists, budding rockpoolers and expert marine biologists alike to learn more, see more and protect our seas for the future.

Sam Fanshawe

Sam Fanshawe – Director, Marine Conservation Society

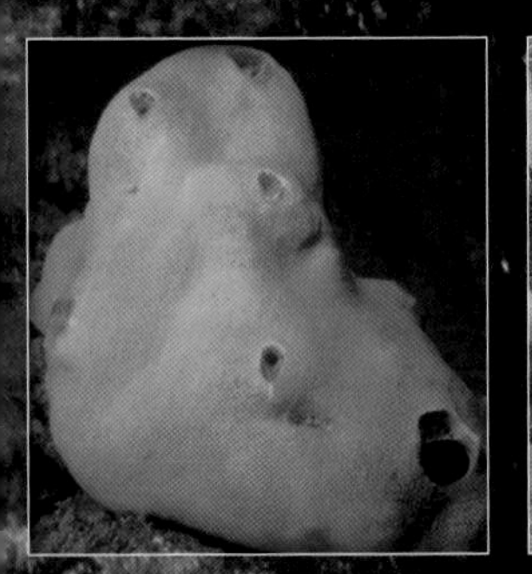

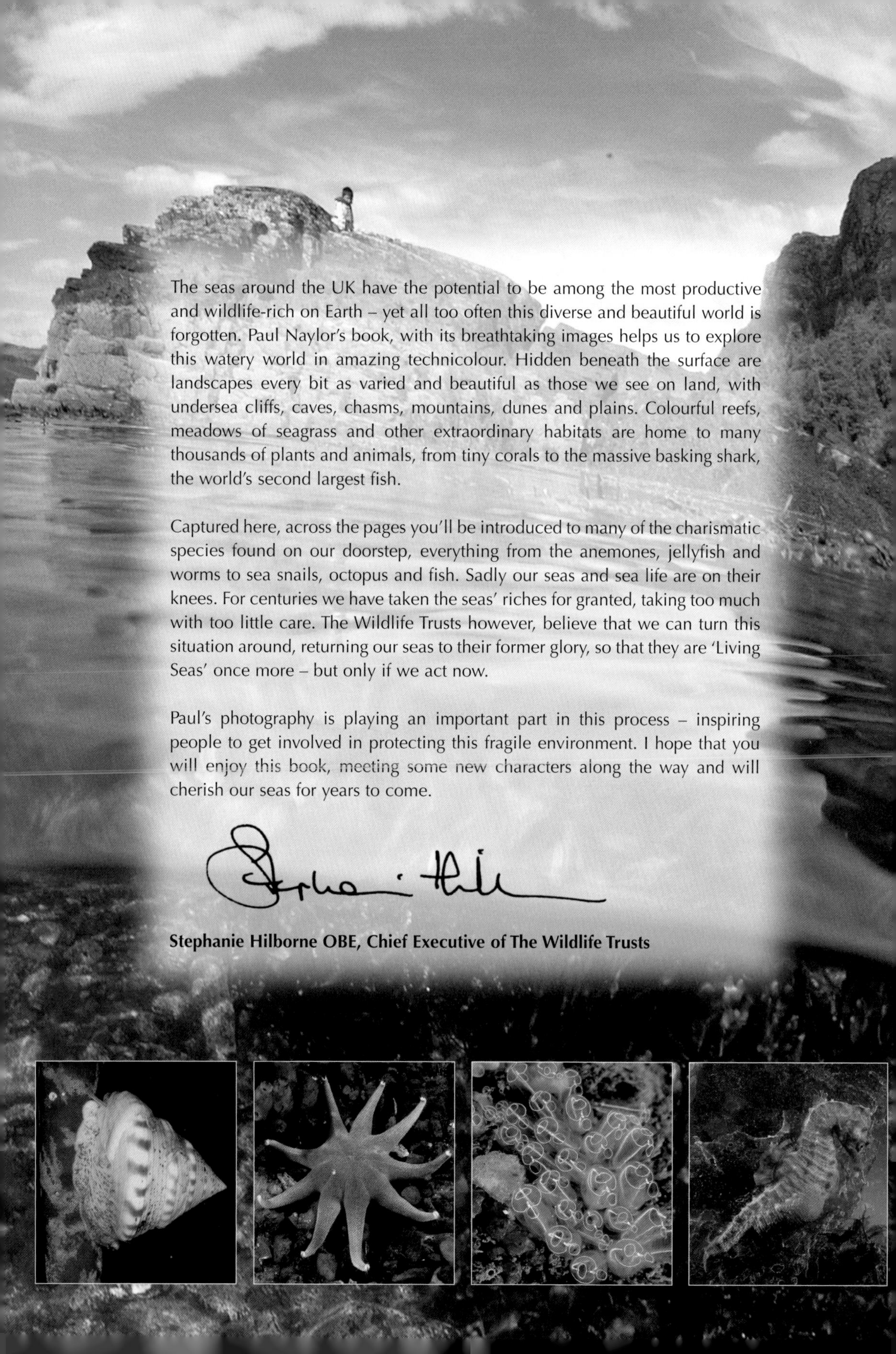

The seas around the UK have the potential to be among the most productive and wildlife-rich on Earth – yet all too often this diverse and beautiful world is forgotten. Paul Naylor's book, with its breathtaking images helps us to explore this watery world in amazing technicolour. Hidden beneath the surface are landscapes every bit as varied and beautiful as those we see on land, with undersea cliffs, caves, chasms, mountains, dunes and plains. Colourful reefs, meadows of seagrass and other extraordinary habitats are home to many thousands of plants and animals, from tiny corals to the massive basking shark, the world's second largest fish.

Captured here, across the pages you'll be introduced to many of the charismatic species found on our doorstep, everything from the anemones, jellyfish and worms to sea snails, octopus and fish. Sadly our seas and sea life are on their knees. For centuries we have taken the seas' riches for granted, taking too much with too little care. The Wildlife Trusts however, believe that we can turn this situation around, returning our seas to their former glory, so that they are 'Living Seas' once more – but only if we act now.

Paul's photography is playing an important part in this process – inspiring people to get involved in protecting this fragile environment. I hope that you will enjoy this book, meeting some new characters along the way and will cherish our seas for years to come.

Stephanie Hilborne OBE, Chief Executive of The Wildlife Trusts

A large ballan wrasse and snakelocks anemones amongst a thongweed forest near Wembury, Devon

Chapter 1
INTRODUCTION
Britain's colourful and fascinating creatures

A wonderful diversity of marine animals live in the sea around Britain. Many people think of their world as "hidden", but these animals can be appreciated while diving, snorkelling, rockpooling or simply walking on the beach. They are staggeringly colourful and beautiful while, even more impressively, creatures which can seem simple at first glance have surprising "tricks up their sleeve" when you examine their behaviour more closely. Beadlet anemones may look like lifeless blobs of jelly at low tide, but they have advanced weaponry for catching prey and blasting their competitors. Limpets appear mundane but regularly navigate their way home across the rocks and find time to intimidate intruders. Those apparently defenceless mussels, if they act together, can defeat a "driller-killer" dog-whelk before it has a chance to eat them.

About this book

Like previous versions, this edition of *Great British Marine Animals* is designed to aid identification of the common animals of British seas, while also including as much information and photographic material as possible showing how they go about their lives. I have produced an expanded third edition because the second edition has sold out and, with a further five years of photography to call on, I preferred to incorporate it rather than produce a straight re-print. This new photography (over 240 images) has enabled the inclusion of thirty more species and, more importantly to me, additional examples of behaviour. These include crabs feeding, sunstars and sand stars spawning, cuttlefish hunting and mating, a sea slug repelling a predator, a commensal ragworm feeding with its hermit crab host and much more...

Spreading the word

Through the efforts of many organisations and individuals, more and more people are becoming aware of the marine wonders around our shores. However, too many people still regard our seas as "grey and uninteresting", and this will only reinforce an attitude of indifference to how we treat them. It is vital that those who are passionate about our marine life continue to "spread the word", and I would very much like to believe that this book can help them in some way. Important progress is being made in marine conservation but there is still much more to do. Organisations that raise awareness of the marine environment and take action to protect it (a selection is listed on pages 312-313) need our support.

Habitats

From the colourful pools of a rocky shore to the muddy but life-filled plains at the bottom of a deep sea loch, a tremendous variety of marine habitats are found in the seas that surround Britain. Each of the panels on the next eight double pages shows a different habitat type and a

Rocky shores and pools

Limpets [page 146], dog-whelk [153]

These are fascinating places for us to explore, but are very challenging habitats for their residents. Animals that live here have to tolerate being left "high and dry" by the tide or, even if they live in pools, withstand wide ranges of salinity and temperature. On many shores they also have to cope with crashing waves and tumbling rocks when hit by storms. These photographs were taken at Wembury, Stoke and Torquay (Devon) and Loch Creran (Argyll).

small range of some of the typical animal species that live there.

The aim of these spreads is to give a flavour of the great diversity and richness in our seas, and show how a selection of habitats support different communities of animals. They are not intended to be a definitive guide to characteristic or key species.

Many factors affect what animals live where, but two of the most important are the type of seabed and the depth of water. It is mainly these that have helped decide the selection of habitats shown here. Seabed type determines how creatures can attach, hide, burrow or move around. Rocky seabeds tend to support the most obvious and colourful forms of life and are therefore usually preferred for snorkelling and diving activities.

Sandy or muddy seabeds can appear desolate by comparison but house many wonderful animals. It is just that many of them spend most of their time out of view beneath the sediment or retreat into burrows when disturbed. Depth of water has a profound effect on habitat characteristics because it determines how much light is available to sustain the growth of seaweed. Light levels diminish rapidly with depth so, as the water gets deeper, animals increasingly dominate the scenery. Below 25 metres, for example, there is not enough light for much seaweed at all.

[Continued on page 14]

Beadlet anemones [page 39], limpets [146], other sea snails

Barnacles [122]

Shore crab [88]

Shanny [290] among barnacles

Dog-whelk [153], limpet [146], top-shell

Snakelocks anemones (small) [42]

Shore clingfish [285]

Rocky reefs in shallow water

Large snakelocks anemones [page 42] on the reef

Ballan wrasse [page 278], rock cook [280], sponges and other encrusting animals

Strong light levels mean the scenery here is often dominated by beautiful beds of kelp, thongweed or other seaweeds. Sponges and other encrusting animals can grow densely where there is shade between the rocks. Numerous mobile animals live among the seaweed and in the rock crevices. All the sights shown here could be seen while snorkelling. The photographs were taken in less than 8 metres of water at high tide, so the areas would be in very shallow water at low tide. They are from Wembury and Babbacombe (Devon), Hythe (Kent), Abercastle (Pembrokeshire) and Loch Carron (Highland).

Rocky reefs in shallow water

Blue-rayed limpets [page 145] on kelp

Corkwing wrasse (male) [282], snakelocks anemones [42], velvet swimming crab [89]

Common and shore sea urchins [220, 222], squat lobster

Sea lemon [170]

Common prawn [119], tompot blenny [286]

Leopard-spotted goby [302]

Velvet swimming crab [89], common mussels [174]

Rocky reefs in deeper water

Jewel anemones [page 60], plumose anemones [44], other sea anemones, dead men's fingers [66], hydroids, ballan wrasse [278]

The scenery here is dominated by animals such as sponges, anemones, corals, hydroids and bryozoans attached to the rocks, as there is limited light to support seaweed growth. Many of the animals are striking and colourful so the views can be spectacular. All these photographs were taken around 20 to 35 metres depth. They are from the Skelligs (Co. Kerry, Ireland), the Manacles and Eddystone Reef (Cornwall), near Plymouth (Devon), St Abbs (Borders), Mull (Argyll) and Loch Carron (Highland).

Rocky reefs in deeper water

Bloody Henry starfish [page 204], dead men's fingers [66], sea loch anemones [58], sea squirts [230]

Cuckoo wrasse (male) [275], sea anemones, corals, sponges

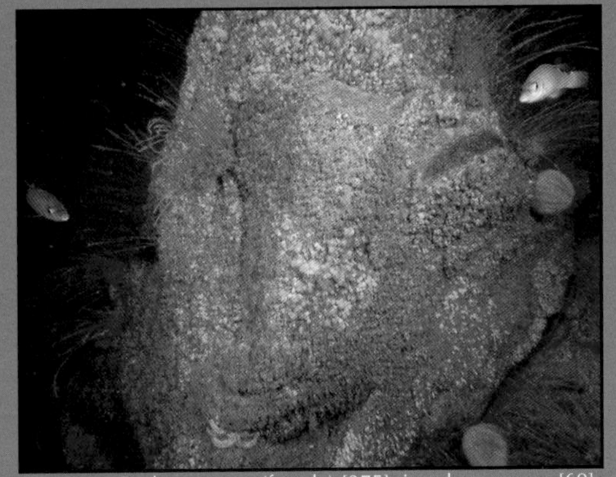
Cuckoo wrasse (female) [275], jewel anemones [60], hydroids, sea urchins [220]

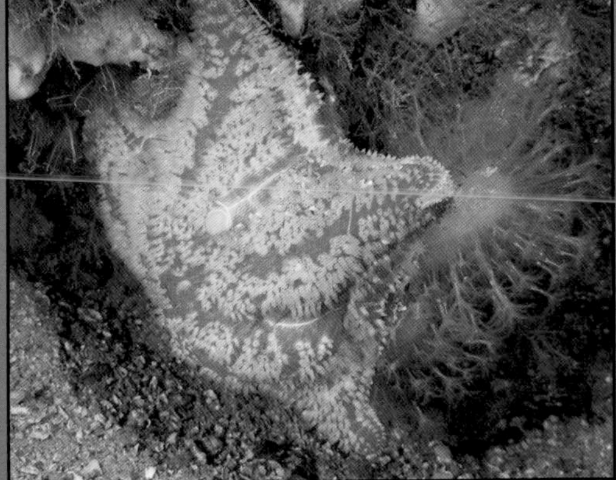
Red cushion star [209] eating soft coral

Ling [256]

Pink sea fans [70], spiny starfish [202], sponges, bryozoans

Sandy seabeds in shallow water

Sand-mason worm tubes [page 132], common hermit crabs [105], lugworm casts [130] and common whelk [156] in background

A large variety of animals make their home here, but can be difficult to see as they are typically well camouflaged or spend much of their time burrowing or buried in the sand. The more time you spend in one place quietly watching, the more likely you are to see the animals appear and get on with their lives. Photographs from Penzance (Cornwall), Torquay and Babbacombe (Devon), Chesil (Dorset), Criccieth (Gwynnedd), Loch Creran (Argyll) and Loch Kishorn (Highland).

Deeper water also protects delicate creatures from the pounding of waves in rough seas.

As well as seabed type and water depth, the degree of water movement (in terms of tidal flow) is very influential because strong tidal flows bring good supplies of planktonic material to the numerous animals that use this source of food. Many of the richest locations for animal life are subject to fierce currents and can only be observed safely during short periods of slack water. The factors affecting what lives where are extremely complex, so please bear this in mind when looking through these introductory pages. Apart from anything else, some species live in many different habitats. I have, for instance, seen velvet swimming crabs and common starfish in all of the habitats portrayed here!

Animal classification

All the following chapters of the book cover individual animal species based on the scientific classification system. Phyla (plural for phylum) are the broadest sub-division of the animal kingdom and make useful groupings. They form, roughly speaking, the chapters. I say "roughly" because animals from more than one phylum are described as worms (Chapter 5), while sea squirts (Chapter 9) and fish (Chapter 10) belong to the same phylum. The order of the chapters has been changed from that in previous editions to reflect current thinking on evolutionary relationships in

Sand star [page 212]

Common cuttlefish [180] digging for sand eels [270]

Common heart urchin [223]

Harbour crab [92] (eating shrimp)

Large necklace shell [152]

Masked crabs (pre-mate pair) [96]

Lesser weever [274]

Sagartiogeton undatus anemone [51], bivalve siphons

Small snakelocks anemones [page 42] on the eel-grass

Formed by eel-grass (a flowering plant, not a seaweed), these undersea meadows grow in some sheltered shallow sandy areas. They provide a particularly rich habitat for a large number of species that hide, feed and breed among the leaves, forming refuges and nurseries for many fish species. Photographs from the Helford Estuary (Cornwall), Plymouth Sound and Torquay (Devon) and Studland (Dorset).

the animal kingdom. In practice, this meant moving the crustaceans to either before the worms or after the bryozoans to keep the right phyla together. For ease, the crustaceans and worms were simply swapped over.

Where they are useful, further sub-divisions of phyla are described in the relevant chapter. I have kept scientific terms to a minimum except for including scientific (Latin) names for each individual species. Scientific species names are undoubtedly useful because some common animals have no English name and others have several, while some English names are applied to more than one species! In this edition, I have added more English names that have recently become better established, mainly from the *Seasearch Observer's Guide to Marine Life of Britain and Ireland* by Chris Wood. Scientific names are unambiguous and precise, although not always constant. The velvet swimming crab, for instance, has had four different scientific names over the last thirty years, while still being called the velvet swimming crab. A few scientific name updates have been included in this edition. A scientific species name comes in two parts: the first part denotes the genus (the narrowest classification sub-division before species) and the second part the species. Very similar animals belong to the same genus and so will have the same first name; totally different animals can have the same second name because it can

Seagrass beds

Two-spot gobies [page 300]

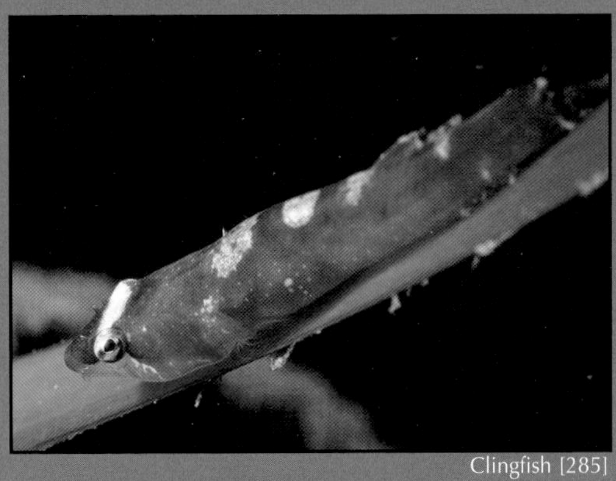

Clingfish [285]

Sea hare [160]

Deep-snouted pipefish [261]

Common hermit crab [105]

Corkwing wrasse (juveniles) [282]

Common cuttlefish [180]

Black goby [298], greater pipefish [260], lugworm casts [130]

Muddy seabeds in deep water

Fireworks anemone [page 58], slender sea pens [72]

These habitats form in deep sheltered water such as in the sea lochs on Scotland's west coast. They might be imagined as unattractive by the uninitiated, but can be eerily beautiful and are the home to spectacular animals such as enormous anemones, elegant sea pens and colourful scampi. Photographs from Loch Melfort (Argyll), Loch Duich and Loch Torridon (Highland).

simply mean "common" or "red" for example. The combination of the two refers to a single species and gives an animal a unique label. The scientific name of the cuckoo wrasse, for example, is *Labrus mixtus* and the closely related ballan wrasse, *Labrus bergylta*, belongs to the same genus.

Using the book

It is very difficult to draw up rules for identifying animals. Key features in one group may be irrelevant in another. The best option is probably to scan through the photographs looking for similar creatures to the one seen, and then refer to the text for more detail. Colour, size and habitat can all give clues to an animal's identity but can also mislead. Information on colour and habitat is given as appropriate, while an idea of approximate size is given for all species in an attempt to put the photographs into perspective. The size given for each species is very much a maximum, so the majority of individuals seen may be considerably smaller. It is also possible, however, that the occasional example is even larger. The reader may find the descriptions accompanied by "often", "usually" and "commonly" frustrating but omitting these words would give a false idea of precision and certainty.

Many of the animals in this book are found all around Britain where there is suitable habitat for them. Notes on distribution of some species are

Muddy seabeds in deep water

Anemone hermit crabs (pre-mate pair) [page 109]

Sand brittle star [218]

Thornback ray [244]

Square crab [99]

Fries's goby [301] by scampi burrow

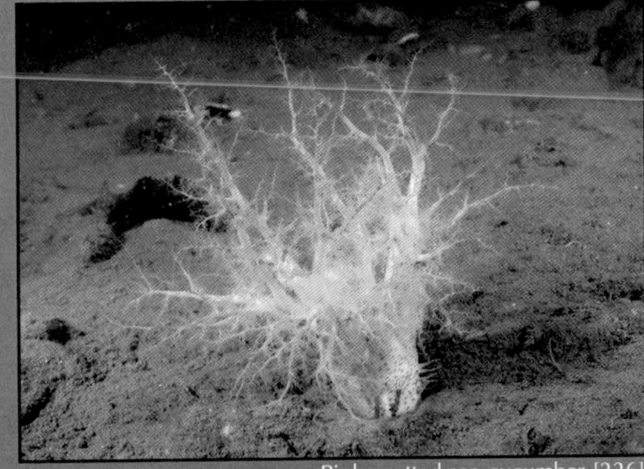
Pink spotted sea cucumber [226]

Scampi [118]

Phosphorescent sea pen [73], scampi in background

Common sunstar [page 205], common sea urchin [220], black brittle stars [217], sea loch anemones [58], peacock worms [135]

Maerl is slow-growing calcified seaweed that forms small, usually pink, free-living nodules. Beds of maerl can form in shallow areas which are sheltered from storms but swept by strong tidal currents, and are rich habitats for a large variety of animal species. Reefs made by file shells can form part of maerl beds and stabilise them to produce a particularly beautiful and vibrant habitat. Photographs from Loch Carron (Highland).

included, with my own observations added where they seem to be relevant. Although the title is "Great British...", there are of course no national borders for marine life and all these animals can be found around Ireland and other adjacent coasts. Where text on a species runs over more than one page, key words are **highlighted** to make information easier to find.

The species covered in this book

This book is far from comprehensive in the species it covers. If expanded to include each of the approximately 7,000 species of marine animals in British seas, it would run to about 25 volumes the size of this one. I have tried to include the most frequently seen and obvious animals, while providing a selection from each group that gives an idea of the types of creatures they contain. More familiar animals are covered in greater detail. The sections on fish, starfish and crabs, for instance, are more wide-ranging (though still far from exhaustive) than those on sponges and sea squirts. In any case, I have deliberately used space showing common and active animals (hermit crabs, cuttlefish, starfish, fish for example) going about their lives in preference to including less familiar and static animals. For more thorough identification of a greater number of species, books with more detail should be consulted (see pages 310-311). Some of these books concentrate on animals within a specific group.

Black brittle stars [page 217] on maerl

Common hermit crabs (pre-mate pair) [105]

Painted goby [299]

Gravel sea cucumber [225]

Sea toad [102] eating gaping file shell [176]

Queen scallop [179], common and black brittle stars [214, 217]

Dahlia anemone [46], sea loch anemones [58], dead men's fingers [66], common brittle stars [214], peacock worms [135]

Chalk gullies and boulders

Edible crab [page 94], encrusting sponges, tube worms

All the photographs here came from an area of chalk gullies and boulders near Sheringham in Norfolk. Although this location shares features with other shallow rocky habitats, the chalk and predominant red seaweed gives it quite a different appearance and crustaceans are especially numerous. Part of the reason for including it is that many people find it astonishing that such colour and beauty lies close to shore in the (often murky) southern North Sea waters.

Finding, watching and photographing Britain's marine animals

Wherever conditions are suitable for snorkelling or diving, or for looking around rocks or in pools on the shore, marine animals can be seen – often in great profusion. Snorkelling in just one or two metres of water near a beach can reveal all sorts of wonders, particularly if there is varied scenery including both rocky and sandy habitats. Look closely at those barnacles; when submerged, their feeding limbs sweep out from the tops of their shells like miniature grasping hands. Small shannies may be watched trying to bite off the barnacles' limbs while keeping a wary eye for hunting fish such as bass sweeping in through the shallows. Crabs can be seen looking around for food too, assuming their defensive claws-spread posture if approached too closely. Around deeper rocks, but still while just snorkelling, beautifully coloured wrasse may be seen building nests and appealing tompot blennies watched defending their territories. Out over the sand, swirling shoals of small sand eels are tracked by a variety of predators such as pollack and larger sand eels. So many interesting aspects of behaviour can be observed with just a little patience. There are descriptions of behaviour throughout the book. Species that give particularly good examples of co-operation, courtship, hunting, camouflage etc. are listed on pages 314-316.

Almost all of the 600 photographs in this book

Chalk gullies and boulders

Olive squat lobster [page 113]

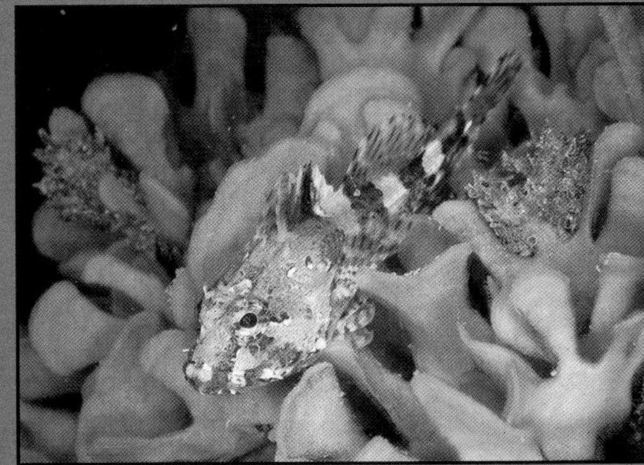
Long-spined sea scorpion [264], hornwrack bryozoan [192]

Common starfish [197]

Velvet swimming crabs (mating pair) [89], edible crab [94]

Edible crabs [94] scraping at the chalk

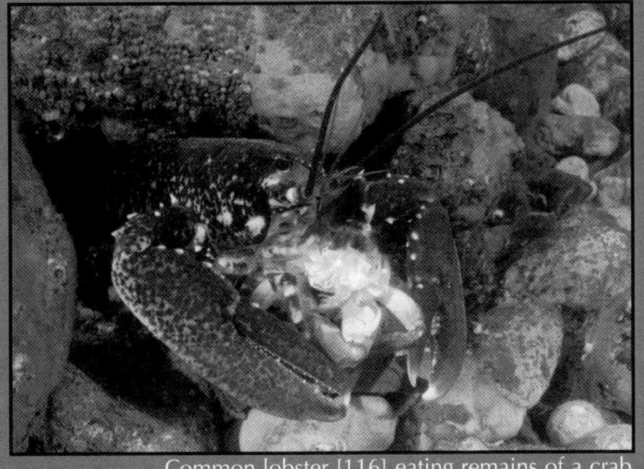
Common lobster [116] eating remains of a crab

Dahlia anemone [46]

Mammals are not covered in this book but encounters, such as with this grey seal, can be magical

were taken underwater, of undisturbed animals in their natural habitat. Fewer than 20 were taken of animals in an aquarium or deliberately uncovered on the shore; and these are labelled as such individually. When taking such photographs, or just enjoying the search for shore life, it is vital that animals are put back and stones or seaweed carefully returned to their original positions as quickly as possible. It is particularly important that large stones are not left upside-down because animals on the undersides need to stay in the damp and dark, and animals on the upper-sides want to be in the light! Underwater, it is essential that divers are careful about where they put their hands, knees and particularly fins. For photographers, it is vital that we avoid damaging fragile marine life when concentrating on getting into a good position for taking that perfect picture. Failure to take enough care makes us the undersea equivalent of aggressive paparazzi!

Photography details

The photographs here were mainly taken on SLR cameras, either a Nikon 801 35mm camera (with Fuji Velvia slide film) or a Nikon D200 digital camera. These were used underwater in dedicated aluminium housings made by Subal. Lenses used included 60 and 105 mm macro, and 20, 16 and 10.5 mm wide-angle. I have also increasingly used wide-angle but close-focussing zoom lenses. Artificial light was used for most photographs, from Sea and Sea or Inon flash guns. All this equipment has given excellent reliability and performance over many years of hard use.

Chapter 2
SPONGES

Sponge diversity

Approximately 400 species of sponge live in the seas around Britain. Just a few of the most obvious, common and distinctive species are included here, and I hope this gives an idea of sponge diversity. Details of an excellent and thorough guide to sponges by Ackers *et al* are given on page 310. This guide makes it clear that microscopic examination is usually essential for definite identification of sponge species.

Simple creatures

Sponges are the simplest members of the animal kingdom. Their cells are specialised for different functions, such as feeding, support or reproduction, but they do not form complex structures like the cells of higher animals. There are no digestive, nervous or circulatory systems for example. The lack of sophistication in sponges has been demonstrated by famous experiments in which, having been broken down by being pushed through fine silk, they soon succeed in reassembling themselves. Because sponges are completely static, they were at one time thought to be plants.

Living filters

Sponges are effectively animated filters and their phylum name, Porifera, means "pore bearer". Water is drawn into the sponge's central cavity through its pores, the numerous tiny holes all over its body surface. Filtered water is expelled via the outlet vents which are much larger, more obvious and fewer in number than the inlet pores. Special cells within the sponge create the water current with continuously beating cilia (tiny whip-like hairs), and collect suspended food particles that are sucked in. This mechanism is simple but effective, and a sponge only a few centimetres across can filter over 20 litres of seawater in a day.

Reproduction

Sponges can reproduce by asexual processes such as budding, or by sexual means. Most sponges are hermaphrodite (simultaneously male and female) but eggs and sperm from the same sponge mature at different times so it does not fertilise itself. Sperm leaves with the outgoing water current, drifts off to fertilise the eggs in other sponges and free-swimming larvae are produced. The larvae then settle and, if the habitat is right, grow into new sponges.

Support and defence

Sponges are supported by a rudimentary skeleton composed of spicules, protein fibres or both. The spicules are needle-like or branched spiky structures made from calcium or silicon compounds. In tropical sponges, it is their fibrous skeleton that produces a traditional bath sponge when the other constituents are stripped away. In addition to providing support, the sharp spicules help to make sponges unpalatable. Many species also produce unpleasant tasting chemicals to deter predators, as any other escape response is beyond their capabilities.

Sea orange - *Suberites ficus*

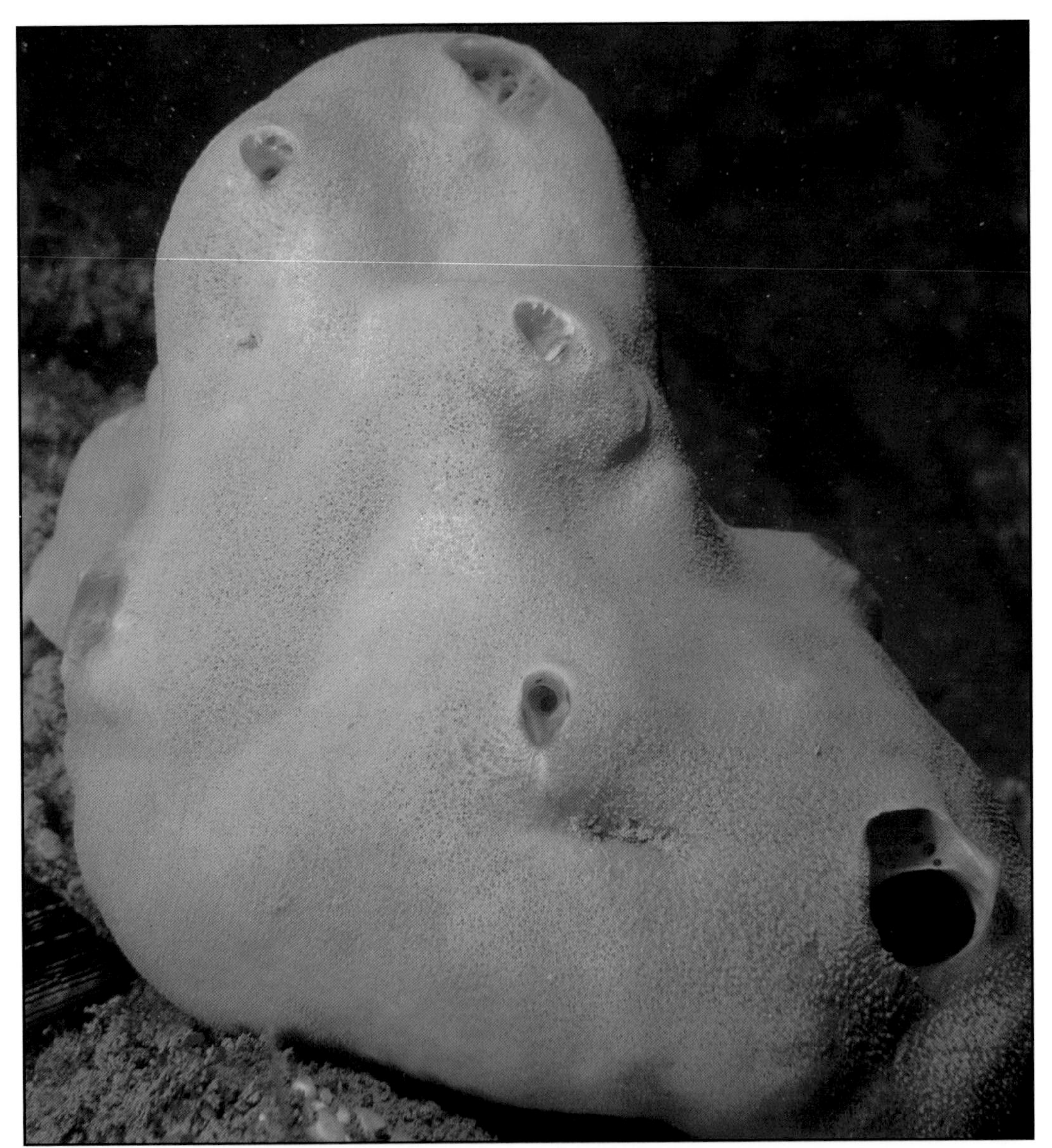

Also known as the sulphur sponge, the sea orange is a classic example of the archetypal sponge. It forms quite large rounded masses that have an even but slightly rough surface, which represents the vast number of tiny pores through which water is pumped into the sponge. In every mass, there is at least one large circular opening through which water is pumped out again. It is easy to peer into one of these openings and see something of the sponge's internal structure. Colour is usually orange but can be brownish or yellow. The sea orange is most common on rocks and stones where there is some mud present. The similar sponge found growing on the shells inhabited by hermit crabs is now considered to be a separate but closely related species. It may completely enclose the hermit's shell and, if this eventually dissolves, the sponge will form the crab's replacement home. [Sea orange up to 20 cm across]

Chimney sponge - *Polymastia penicillus*

At first sight, this species does not really look like a typical sponge and could almost be mistaken for a colonial sea squirt. It is often found in quite silty locations and the sponge's cushion-like base may be obscured by a covering of sediment so only a forest of projections is visible, as in the photograph. Some of these projections are narrow and have pointed tips while others (the outlet vents) are wider open-ended tubes like miniature chimneys; all are usually creamy-white and translucent. [Base of sponge up to 15 cm across, "chimneys" up to 10 cm tall but usually far smaller]

Hedgehog sponge - *Polymastia boletiformis*

This species is similar to the closely related chimney sponge (above) in that it has a cushion-like base with many projections extending upwards from it. However, the hedgehog sponge's cushion is likely to be prominent and visible, rather than hidden by silt like the flatter chimney sponge's cushion. The projections on the hedgehog sponge are thicker and it is generally brighter yellow or orange in colour. The water outlet openings at the ends of some of the projections will often be wide open, as on the sponge in this photograph. [Up to 10 cm across]

Boring sponge - *Cliona celata*

Extensive growth of boring sponge (surrounded by white dead men's fingers) on a reef near the Lizard, Cornwall

The name of this sponge arises, not from its uninteresting nature, but from the fact that it bores its way into soft rock such as limestone. In many instances, most of the sponge is hidden within a network of passages and chambers that it has excavated in the substrate. All that is visible on the seabed are then the characteristic yellow "studs" and vents (see small photograph) where water enters and leaves the sponge respectively. Sometimes, however, the sponge outgrows its chambers and can form large, very obvious masses, which are still covered in the familiar "studs" and vents. The main photograph shows a particularly extensive growth. The sponge's boring process is chemical, and employs an acid that is a by-product of its respiration. Special cells use this process to undercut and surround tiny pieces of rock which are then "spat out" with the water flow. The sponge also bores into the shells of molluscs and can be a serious pest in commercial oyster beds. [Masses can reach up to 1 m across but are usually much smaller]

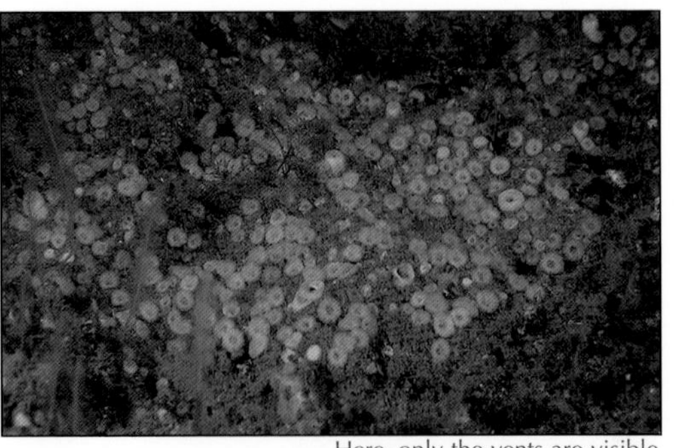

Here, only the vents are visible

Golf ball sponge - *Tethya citrina*

This distinctive sponge is indeed shaped like a ball and is covered with bumps, some of which form bud-like projections. Its short stem, which is attached to the rocky seabed, is usually hidden from view by the rest of the sponge. A single large water outlet hole can be seen in the top of the ball, opposite the attachment point. Colour is generally yellow or orange, and there is often a dusting of silt. The whole sponge can shrink down to less than half its full size when it is disturbed. [Ball up to 7 cm across]

Chocolate finger sponge - *Raspailia ramosa*

Colonies of this sponge resemble small dense bushes, with the "branches" having rounded and slightly bulbous ends. Its colouration is an attractive dark reddish-brown that can appear almost crimson under torch-light. However, fine silt often adheres to its velvety surface layer and obscures this rich colour. Small water outlet openings can be seen scattered across the surface of the branches. ["Bushes" up to 15 cm tall]

Yellow staghorn sponge - *Axinella dissimilis*

A distinctively shaped sponge which is usually found in quite deep and clear water offshore. Its orange or yellow fan-like form stands erect from rock faces. The branches, which may be joined for part of their length, are flattened and therefore oval in cross-section, with rounded ends. The overall shape is slightly reminiscent of the sea fan (page 70). While totally different types of animal, their form presumably serves the same function in both cases: to maximise the surface area that comes into contact with food-bearing currents. [Up to 15 cm tall]

Prawn cracker sponge - *Axinella infundibuliformis*

This off-white or beige sponge has a very characteristic form and is usually shaped like a funnel or shallow wine glass. The cup is attached to the rock, on which it lives, by a short stalk or stem. The rim of this cup is thick and rounded. The surface of the sponge is fairly smooth and the numerous water outlet holes spread evenly across it are clearly visible. Even though it is often found on quite silty rocks, the sponge's cup generally appears clean. White and orange dead men's fingers (pages 66-67) can be seen beside the sponge in the photograph. [Cup up to 25 cm across but often much smaller]

Breadcrumb sponge - *Halichondria panicea*

A very common encrusting sponge that can form large sheets or lumps on rocks in shallow water, often beneath overhangs. The sheets may be thin or quite thick, with the raised water outlet holes looking like miniature volcanoes or chimneys. It can occur in a variety of shades, from olive green through a dirty cream to pale yellow. The green colouration is due to algae which live symbiotically within the sponge's tissues, and are dependent on the amount of light available. Breadcrumb sponge found under gloomy overhangs or in deeper water has less algae and is therefore more yellow. [Encrustations can reach over 1 m across and are of very variable thickness]

Red encrusting sponge - *Clathria atrasanguinea*

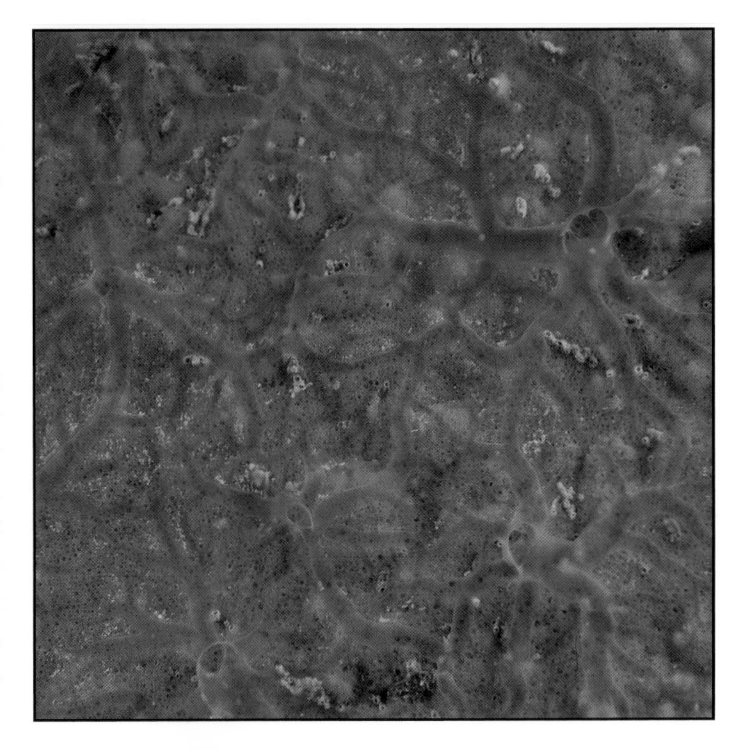

The brilliant deep red colour is a common feature of this sponge and the large patches it forms on rock faces, wrecks or pier legs look like splashes of red paint. These encrustations are only a few millimetres thick and are fairly smooth in comparison with many sponges. Wavy branching channels in the surface of the sponge converge into water outlet openings to produce a distinctive pattern, but identification cannot be certain without microscopic examination of the sponge's spicules. [Encrustations up to 30 cm across]

Honeycomb sponge - *Hemimycale columella*

Often known as the "crater sponge", the surface of this pale pink, orange or yellow encrusting sponge is indeed covered with characteristic craters. The rims of the craters are particularly obvious because they are paler in shade than the rest of the sponge and this produces the honeycomb pattern. Each of the craters contains one or more small but visible water outlet holes. This sponge is most common in the south west of Britain. [Encrustations up to 1 cm thick and 30 cm across]

Shredded carrot sponge - *Amphilectus fucorum*

This species is an encrusting sponge that can form sheets or mounds. It is usually deep orange in colour, and has a delicate, quite flexible consistency. Under more sheltered conditions, it forms many distinctive long and slender tassels that stick out from the rest of the encrustation. The large water outlet holes may be scattered over a flat surface of the sponge or raised up like chimneys. It grows with great abundance in many locations such as the Menai Straits and Plymouth Sound. [Encrustations of very variable size, tassels can be 10 cm long]

Mermaid's glove - *Haliclona oculata*

This sponge forms characteristic shrub-like colonies which can grow upwards from a flat seabed or stick out from a vertical rock surface. The branches, which are often numerous, are dirty yellow or beige in colour, have a round cross-section and bear distinct water outlet holes along their length. This species likes tide-swept areas but can tolerate quite silty conditions and is common in the outer reaches of estuaries. ["Shrubs" can be up to 30 cm tall, branches 1 cm across]

Volcano sponge - *Haliclona viscosa*

The purple to pink colour of this sponge is reasonably distinctive, as are the prominent water outlet openings. These often look like miniature chimneys or volcanoes (hence the name) but they can also merge to form ridges as on the sponge in this photograph, taken near the Lizard in Cornwall. The rock around the sponge is covered with a "turf" made up of seaweed and of animals such as hydroids and bryozoans. There are also white dead men's fingers (page 66). [Sponge up to 30 cm across]

Elephant hide sponge - *Pachymatisma johnstonia*

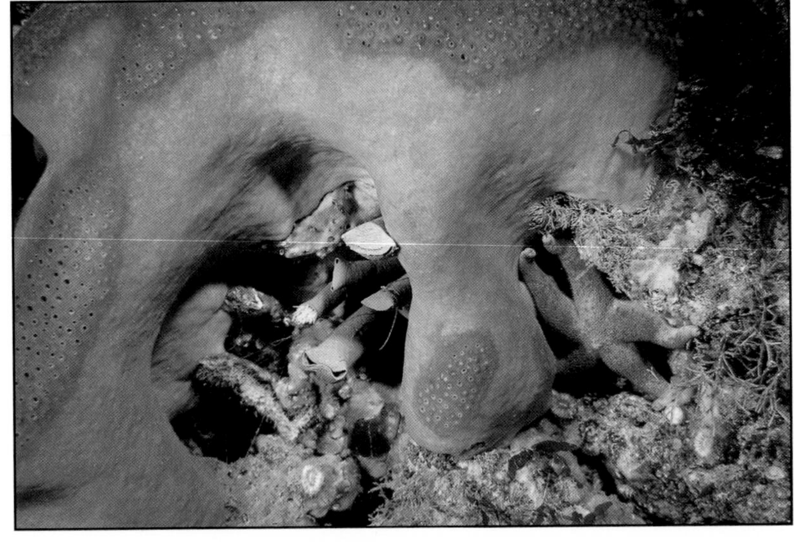

The large grey mounds or plates formed by the elephant hide sponge (also known as the elephant's ear sponge) are usually found protruding from vertical rock faces. Its surfaces are smooth and hard with obvious holes arranged in patches or lines, where water that has passed through the sponge is expelled. Water is taken in through more numerous and much smaller openings all over the body of the sponge. This photograph shows a typical large growth of the sponge; close to it on the reef are fan worm tubes (page 134), a bloody Henry starfish (page 204) and Devonshire cup-corals (page 64). Elephant hide sponge can sometimes be coloured blue or white. [Plates of sponge can be over 1 m across but are usually smaller]

Goosebump sponge - *Dysidea fragilis*

The common name of this abundant encrusting species is a very good one as the small conical projections that cover its surface look just like "goose bumps". The sponge forms rough sheets or small mounds made up of several lobes. Circular water outlet holes are scattered over its surface. Colour is cream or grey, occasionally mauve or brown. There are no silicon or calcium-based spicules in this sponge's skeleton, which is made just from protein fibres. [Growths up to 30 cm across]

Goosebump sponge among other sponges, bryozoans and hydroids

Purse sponge - *Sycon ciliatum*

A species with a roughly cylindrical shape, rather like a tall, slim vase. The dense cream or brown coloured hairs that cover it produce a distinctive shaggy appearance. It can be found attached to rocks or seaweed, often in small groups. The single water outlet hole is positioned at the free end of the sponge and is surrounded by a ring of longer, stiff hairs. The body wall is supported by calcium-based spicules and it is classified in a different group from the preceding species in this chapter, most of which have silicon-based spicules. [Up to 4 cm long]

White lace sponge - *Clathrina coriacea*

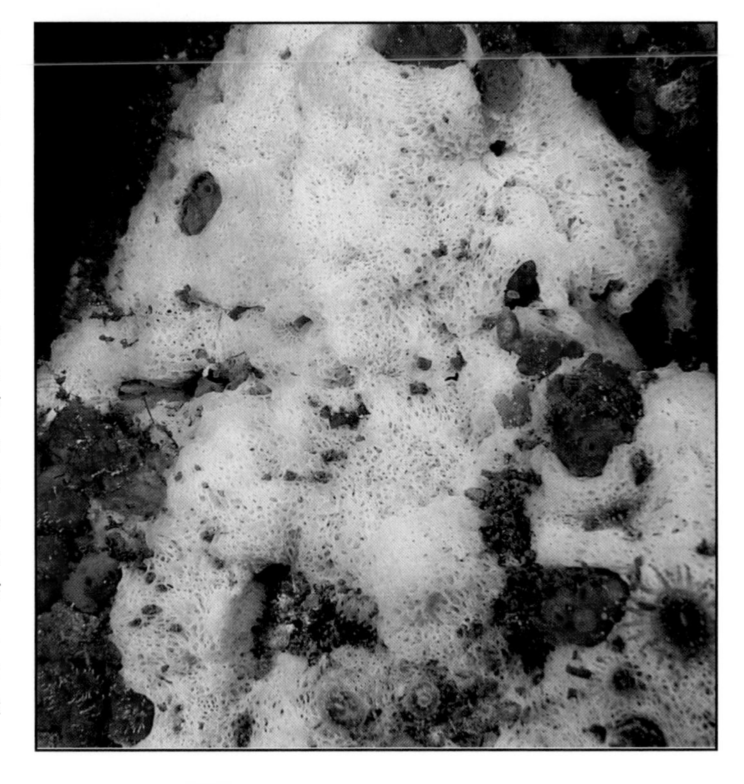

This sponge has a distinctively mesh-like appearance, as it consists of a mass of delicate interconnecting tubes. Usually white in colour, it forms a very obvious encrusting sheet over rock surfaces. It is often found living with the red gooseberry sea squirt (see page 236) as it is in this photograph. The water inlet and outlet holes of this species are too small to see with the naked eye. Like *Sycon ciliatum* (above), the skeleton of this sponge is based on compounds of calcium rather than silicon. [Encrustations up to 1 cm thick; irregular in outline but up to 20 cm across]

Chapter 3
CNIDARIANS
Sea anemones, corals, hydroids & jellyfish

Armed and dangerous

The distinguishing feature of cnidarians is that they have stinging cells. These cells contain discharge capsules, called cnidae, that give the group its name. The discharge capsules, which are used for both defence and the capture of prey, are impressive examples of engineering in miniature. Each capsule contains a long hollow, coiled thread which uncoils and shoots out under water pressure when the cell is triggered by touch or chemical stimulus. Different threads have varied functions and, when thousands are triggered together, they can have a powerful effect. Some simply entangle the prey, while others stick to it or inject poison. Some even have blades that, in combination with the twisting action of the threads, act as tiny drills on the armoured surfaces of small crustaceans.

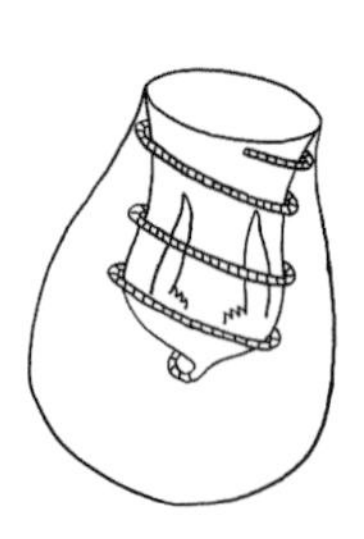

A discharge capsule (cnida) before and after discharge.

A step up from sponges

Apart from their exceptional weapon system, cnidarians are fairly simple animals. They have different tissues specialised for various functions, so they are a step up from the sponges, but they do not have proper organs like higher invertebrates. There is no true circulation system and only an extremely simple nerve network. Their tentacles, covered in dense batteries of stinging cells, capture prey animals and pass them to the central mouth where they are engulfed - see photographs (opposite) of a beadlet anemone swallowing a shrimp that it has caught. There is no anus so the mouth is also used for the expulsion of undigested material.

Polyps and medusae – their role in the different groups

Cnidarians can occur in the form of either a polyp, living anchored to the seabed, or a free-swimming medusa.

Sea anemones only occur in the polyp form. These flower-like animals are almost always found attached to rocks or other hard surfaces. Water pressure inside the body maintains the anemone's shape and provides a base for muscle action. They reproduce by producing eggs which normally develop into new adults via a

The beadlet anemone at the top is eating a shrimp

......a few minutes later (aquarium photographs)

planktonic (floating) larval stage. Asexual reproduction may also occur where an adult anemone splits or buds to form a new individual and dense colonies can result.

The following terms are often used when describing sea anemones:

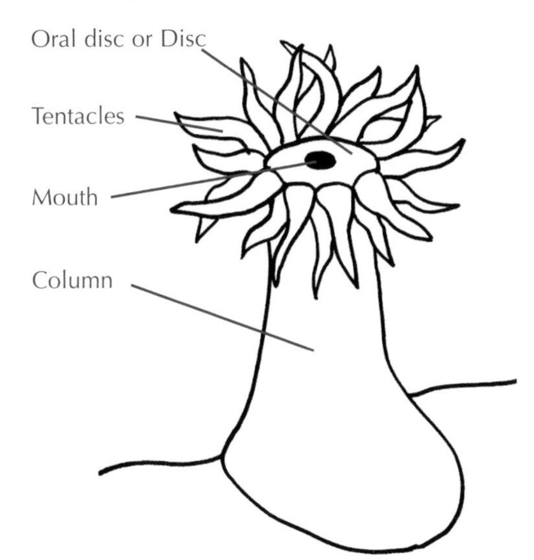

Corals are essentially the same type of animal as sea anemones, occurring only in the polyp form, but they produce some sort of skeleton to support and protect their bodies. Stony corals are usually colonial, their fused chalky skeletons able to form huge coral reefs in tropical waters. Some, including the commonest British species, are completely solitary. In terms of biological classification, the sea anemones and stony corals are very similar and belong to the same sub-group. Within this sub-group there are further anemone-like species such as the burrowing anemones, the encrusting anemones and the jewel anemone which are not classified as "true" anemones. Soft corals, sea fans and sea pens belong to a related but separate sub-group. They are colonial animals whose "mutual" skeleton is not a solid mass of calcium carbonate, but is gelatinous with embedded calcareous or horny spicules imparting strength and some rigidity.

Hydroids, or sea-firs, are the simplest of the stinging-celled animals. Most hydroid species occur in both the polyp and medusa form at different times during their life cycle, but the polyp stage is dominant. The medusa stage may be free swimming or be simply an extension to the polyps. The polyps form colonies where they are linked by strands of living tissue and individuals may serve different functions within the colony. Some polyps, for example, are responsible for feeding while others form the reproductive medusae.

Jellyfish are like hydroids, in usually having both polyp and medusa stages, but it is the medusa stage that dominates as a large floating predator. Its body typically forms a "bell" or "umbrella" which can contract rhythmically to propel the animal through the water. As in all cnidarians, tentacles armed with stinging cells catch prey and pass it to the mouth, so jellyfish can be visualised as floating sea anemones. The diagram below shows a typical jellyfish life cycle. Male jellyfish release sperm into the seawater which is drawn in through the female's mouth and fertilises her eggs. The eggs then develop within the female until swimming larvae are produced. These larvae settle on the seabed and develop into small polyps with long tentacles. The polyps grow and split across so their bodies eventually become a stack of saucer-shaped structures. The "saucers" separate off as miniature jellyfish and grow into adults. All the jellyfish species described in this book follow a similar life cycle, except for the mauve stinger (*Pelagia noctiluca)* which lacks a bottom-living polyp stage. [Note that the comb jellies on page 83 belong to a different phylum from all the other species in this chapter]

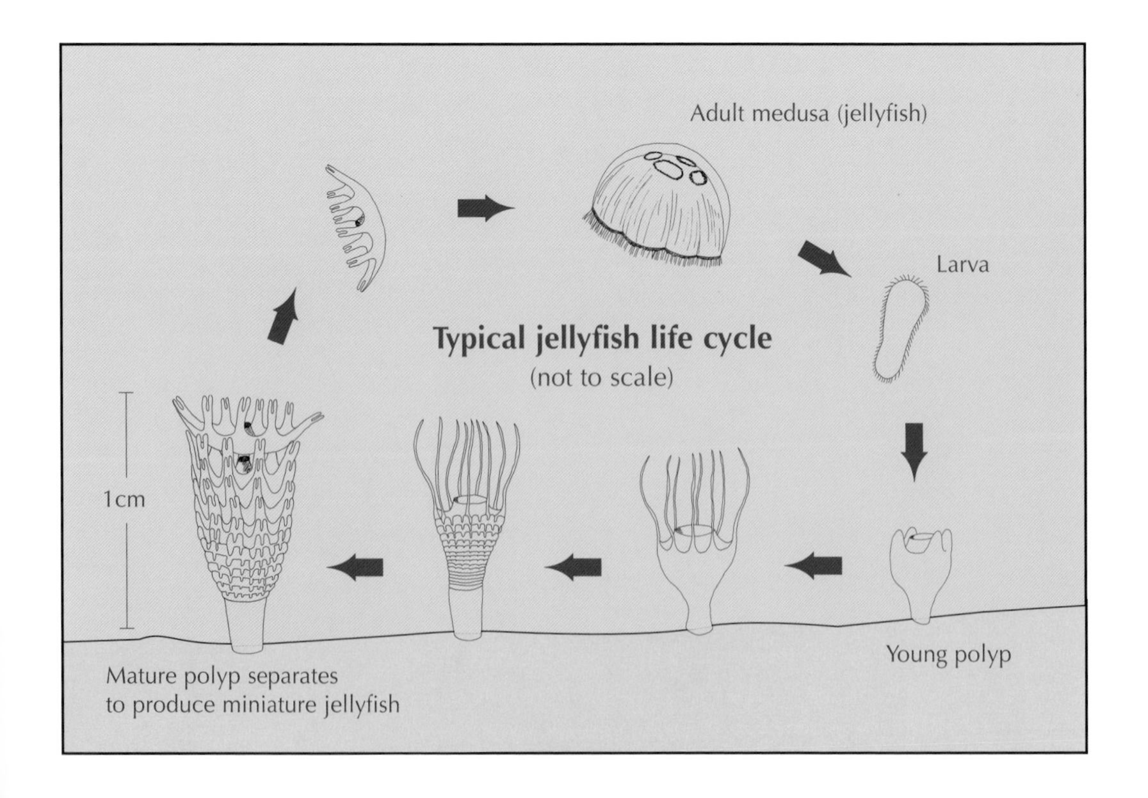

Beadlet anemone - *Actinia equina*

The beadlet anemone is usually seen on rocky shores or in very shallow water. Underwater, its dense mass of tapering tentacles can be fully appreciated (see above and page 40 top right) but when exposed to air at low tide it retracts to avoid dehydration and then resembles a blob of jelly (page 40 top left). Beadlet anemones are most often coloured a deep red but can be green, brown or orange. There is increasing evidence, however, that some of these colour forms may be separate species. The beadlet anemone's column is smooth but there may be small blue bulges,

Beadlet anemone - *Actinia equina*

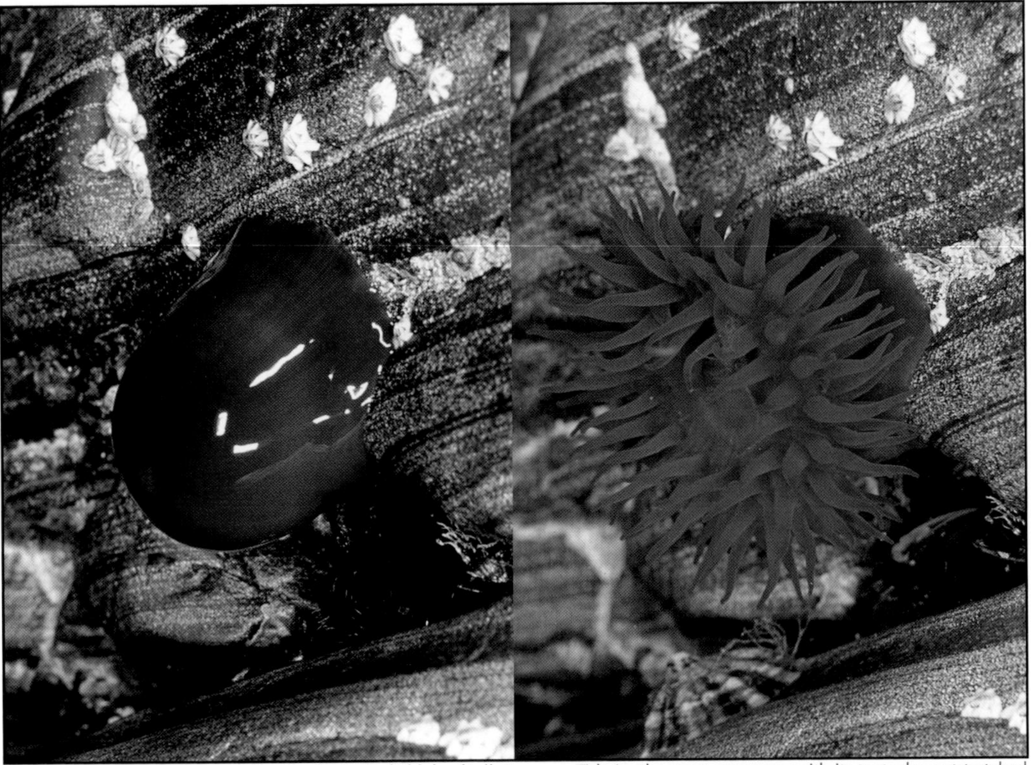

Tide out: the typical "blob of jelly"

Tide in: the same anemone with its tentacles outstretched

known as tubercles or acrorhagi, visible in a ring just below the tentacles (see below). These acrorhagi, which contain large numbers of stinging cells, are important in aggression.

Acrorhagi (blue bulges) revealed as the tentacles move in the swell

Beadlet anemones fight each other as they compete for favoured spots on the rocks. When two anemones meet, their tentacles touch but are then withdrawn from the area of contact. The acrorhagi (blue bulges) are the offensive weapons and, by bending towards their opponent, the aggressive anemones bring them into play. The bulges swell up before firing off a barrage of stinging harpoons and their surface may then peel off and remain attached to the target where it continues to cause damage. Such a bout ends when one of the combatants retreats. Unlike many sea anemones which produce planktonic larvae, the female beadlet broods its embryos for a few weeks before releasing them. Asexual reproduction also occurs, with internal budding seemingly producing miniature new anemones that develop inside both male and female adults. [Up to 5 cm across]

Strawberry anemone - *Actinia fragacea*

The column of this dark red anemone bears characteristic bright green spots, so the derivation of its name is obvious. The strawberry anemone has a very similar shape and form to the beadlet anemone but is larger than its close relative and is found in slightly deeper water. For a long time classified as simply a different colour form of the beadlet, it is now recognised as a separate species. Unlike the beadlet, it does not appear to brood its young. [Up to 10 cm across]

Gem (or wartlet) anemone - *Aulactinia verrucosa*

Despite the similarity of its name, this small and pretty anemone is a totally different species from the jewel anemone (pages 60-61). Often found in lower shore rock pools, it is mainly restricted to western and south western British coasts. The usual colouration is a blend of pink and grey with green markings around the mouth. The relatively few (48 or so) tentacles are translucent with attractive opaque spots. A few of the "warts" or "gems" that give rise to the anemone's common names are visible around the edge of its disc in the photograph. These warts, which are arranged in rows (some white and others dark) on the column are much more obvious when the anemone is closed. When tightly contracted, they make the anemone look like a small sea urchin skeleton without the spines. Young gem anemones are brooded within the parent. [Up to 5 cm across]

Snakelocks anemone - *Anemonia viridis*

The snakelocks anemone prefers the brightly lit, seaweed-rich areas of rocky reefs in shallow water (top photograph this page) to the darker world of cliffs and overhangs which many anemone species inhabit. It can also be found in pools on the shore and on kelp fronds or, in the case of small individuals, on eel-grass strands. The two hundred or so long wavy tentacles are very sticky and are often a rich green colour with beautiful purple tips; they are unusual in that they cannot be fully retracted. The tentacles may obscure the short and squat column which is tapered and irregular in cross-section. The sun-seeking habit of the snakelocks is related to the fact that the tissues of its tentacles contain large populations of special symbiotic algae. In this close association, the algae gain protection and a supply of carbon dioxide and nutrient salts; the host anemone benefits from organic compounds synthesised by the algae using sunlight as the energy source. The algae may also help to remove waste products from its host's tissues. Snakelocks in deep or murky water are often a dull grey colour and this may be due to their algal populations becoming depleted in low light conditions. This species of anemone also has some sort of relationship with Leach's spider crab, *Inachus phalangium* (see page 104). The crabs seem content to reside around the base of smaller anemones but, when a larger anemone is their home (bottom photograph), they can be found living right in the centre. The snakelocks is absent from most of the east coast of Britain. [With its long tentacles, can be up to 20 cm across]

Snakelocks among seaweed on top of a reef

Small spider crab residing in snakelocks

Snakelocks anemone - *Anemonia viridis*

Plumose anemone - *Metridium senile*

A group of these prominent animals swaying in the current makes a spectacular sight. When fully active, the plumose anemone has a tall smooth column topped with a crown of very numerous fine, slender tentacles which give the characteristic feathery appearance. When withdrawn, it appears as a contracted little mound (see front right of photograph above). Individuals may be white, orange, green or brown. Tentacles are usually, but not always, the same colour as the column (see above). Plumose anemones show a marked preference for areas of strong water flow, so they are often found on rock pinnacles or prominent pieces of wreckage. While normally most obvious below the seaweed zone, plumose anemones can be seen in shallow shady spots such as under overhangs and on jetty pilings. With fine delicate tentacles, these anemones specialise in capturing very small prey and their enzyme secretions can break down the shells of small planktonic crustaceans. As well as flexing whole tentacles to pass prey to the mouth, the plumose can use hair-like flagellae and mucus strings on each tentacle to transfer food there. The results of asexual reproduction by these anemones can often be seen, where parts of the base break off to form new anemones (see photograph left). [Up to 30 cm tall]

Small new anemones form from pieces that break off the base

Plumose anemone - *Metridium senile*

Dahlia anemone - *Urticina felina*

The dahlia anemone has a sturdy appearance, with its short squat column covered in warts and its rather stout tentacles, but it is nevertheless a beautiful animal. A powerful predator, it can catch and devour active prey such as prawns and surprisingly large fish. Dahlias can occur in a variety of different colours (see photographs) and often have attractive banding on their tentacles as well as radiating patterns on the large oral disc. Gravel or shell fragments are usually stuck to the column so that, when the anemone is fully retracted, it is surprisingly inconspicuous. Dahlias are often seen

Dahlia anemone - *Urticina felina*

singly but can also occur in dense patches at the bottom of shallow rocky gullies. Such aggregations are a marvellous sight to the passing diver, but must spell doom to many an unwary small fish. The surge conditions found in such gullies will of course make it more difficult for the anemones' prey to avoid the grasp of their tentacles. Given the dahlia anemone's **predatory and voracious nature**, I was surprised to observe the sequence of events shown in the photographs on this page. A small group of painted gobies (page 299) seemed simply, at first, to be straying perilously near to the outstretched tentacles of a medium-sized anemone. The gobies, however, then appeared to be deliberately disturbing sand from the seabed so it drifted over the anemone's tentacles. At least one goby seemed to bite a tentacle. After a few moments of this attention, the dahlia anemone partially retracted its tentacles and the gobies then seemed to ignore it and move away. I have not heard of any reference to similar behaviour by fish towards anemones elsewhere, but could it be explained as deliberate group action by the gobies to reduce the threat from a potential predator? [Up to 20 cm across]

A painted goby ventures very close to a dahlia anemone

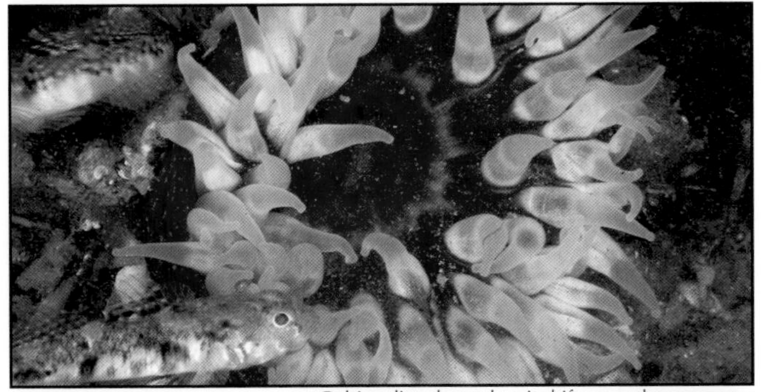
Gobies disturb sand so it drifts over the anemone

A goby (bottom left) appears to bite a tentacle

The anemone partially closes up

Dahlia anemone - *Urticina felina*

Dahlia anemones in the Menai Straits

Trumpet anemone - *Aiptasia mutabilis*

The trumpet anemone is essentially a Mediterranean species that, in Britain, is limited to the far south west. Here, it can be quite common, large numbers are found in Torbay for instance, but its kelp-like colouring means that it may stand out less than other anemones. The overall brown or khaki is broken up by distinctive white or pale blue lines on its disc which radiate out from the mouth. [Can apparently reach up to 15 cm across but usually 5 cm or less]

White striped anemone - *Actinothoe sphyrodeta*

A very common small anemone on southern and western coasts, it can be found individually or in groups on rock faces. Most individuals are white all over but the disc is sometimes orange. The column and tentacles are always white. There are usually faint dark vertical stripes on the column, particularly visible when the anemone is contracted. This species can be confused with the all-white or white-orange forms of *Sagartia elegans* (see page 50) but *Actinothoe* has untidier-looking tentacles which are fewer in number, and no suckers on its column. It is sometimes found attached to the large sea squirt, *Phallusia mammillata*, see page 234. [Up to 2 cm across]

Elegant anemone - *Sagartia elegans*

Individuals of the form with patterned disc and tentacles

This anemone can be found in several different colour forms which, at first glance, look like separate species. The main photograph shows a group of these anemones of the form with patterned disc and tentacles; it was taken beneath the famous arch of Cathedral Rock at St Abbs (south east Scotland). The smaller photograph, taken in Cornwall, shows individuals of the following forms: white tentacles-white disc, white tentacles-orange disc and pink tentacles-variable coloured disc. A further permutation, with orange tentacles and a variable coloured disc, is quite rare and makes five colour forms in all. Regardless of tentacle and disc colour, the anemone's column is usually dull orange and is covered with small wart-like suckers that are most obvious when the tentacles are withdrawn. Sticky white threads are sometimes released by the anemone if disturbed. Reproduction can either be sexual or by an asexual process where fragments of tissue separate from the base and form new anemones. [Up to 5 cm across]

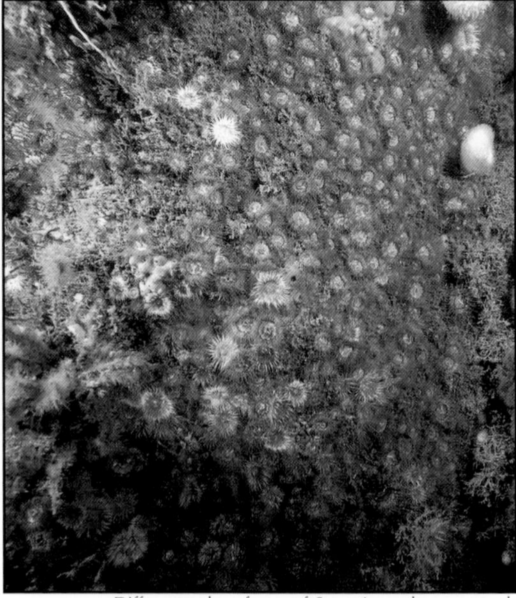

Different colour forms of *Sagartia* on the same rock

Sagartiogeton undatus

An attractive but relatively unobtrusive anemone that can be quite common in sandy areas. Its base is usually attached to a stone or shell buried beneath the surface of the sand, and an anemone is often found right next to where the siphons of a bivalve mollusc reach the sand's surface (see small photograph). The anemone's column can be tall enough to stand well clear of the seabed when it is fully extended (see main photograph). The column, an overall yellowish brown, has faint stripes running up it and there may be dark spots near the top. The disc has an attractive pattern, usually in grey and cream, while the long slender tentacles are translucent. There are two thin dark lines running down each tentacle, a useful distinguishing feature. The whole anemone can contract down into a very small mound when disturbed. [Up to 12 cm tall]

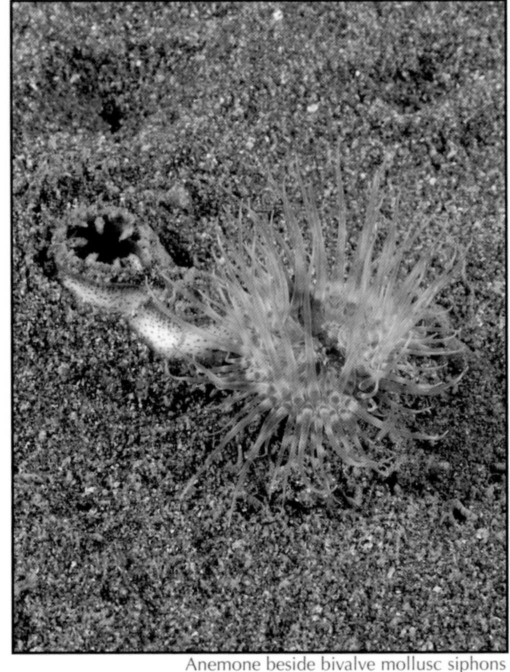

Anemone beside bivalve mollusc siphons

Sagartiogeton laceratus

This species can perhaps best be described as a small, brightly coloured version of *Sagartiogeton undatus* (page 51). It can be found attached to rocks, shells, worm tubes or stones, sometimes partly buried in sand. The column is usually pale orange and there is also plenty of orange on the attractively patterned disc. Pale V-shaped stripes stand out from darker markings around the bases of the slender tentacles. The tentacles have no dark lines running down them (as do those of *S. undatus*). As indicated by the second part of its scientific name, *S. laceratus* undergoes asexual reproduction where parts of its base break away (by laceration) and form new anemones (see photograph). This is why the edge of the anemone's base tends to be very irregular in shape and several individuals are often found living together. [Up to 6 cm tall]

Daisy anemone - *Cereus pedunculatus*

Can form carpets of many individuals or be found singly, often living on muddy seabeds where their bases are anchored to stones buried in the sediment. The anemone's long, slender column is hidden in the mud so its large disc and numerous short tentacles lie virtually flush with the mud's surface. Daisy anemones also live on rock where their columns are hidden in crevices. The disc may be a uniform brown or be attractively patterned in various colours, while the tentacles can be striped or mottled. There is often a bold splash of colour around the mouth. Young anemones develop within the parent and are released as fully formed miniatures. [Up to 10 cm across, more usually about 5 cm]

Peachia cylindrica

This species has a distinctive appearance with 12 tentacles (much fewer than most anemones) that are mottled in brown, grey and white with repeating W-shaped markings down their length. Its disc can be patterned with similar colours or be all white. The anemone lives in sandy habitats and the tentacles are usually seen lying out flat across the seabed while its long slender column is buried in the sand. This column has a rounded end (rather than the adhesive base of most anemones) and acts as an effective anchor which stops the anemone being washed away by currents and swell. [Tentacle span up to 15 cm across but usually much smaller, buried column up to 30 cm long]

Parasitic anemone - *Calliactis parasitica*

Common in the south west but rare further north, this species of anemone is usually found on second-hand mollusc shells inhabited by the large hermit crab, *Pagurus bernhardus* (pages 105-108). The column of the parasitic anemone is typically a dirty cream colour with brown spots or stripes, while the tentacles are a yellowish grey. The anemone's name is misleading because it is in no way a parasite. Both crab and anemone benefit from the relationship and both can live independently; the anemone sometimes being found on stones or empty mollusc shells (top photograph, opposite page). Such a loose liaison is not true symbiosis either, and is normally referred to as commensalism. When living together, the crab will benefit from extra protection provided by the anemone's stinging tentacles against predators such as fish and cuttlefish. In return, the anemone receives extra scraps as the crab messily rips up its food, and gets free transport to different feeding locations. As the crab moves around, the anemone will often bend over so its tentacles "sweep" the seabed. Parasitic anemones can recognise hermit crab shells, probably by smell, and will actively transfer onto one if they are living on another substrate. First of all, they reach out and attach their oral disc to the shell with the aid of discharge capsules. They then detach their base from its current anchorage, and swing it across onto the hermit's shell. The crab plays no active part, but co-operates by keeping still. Occasionally, more than one parasitic anemone will be found on a single hermit crab. The bottom photograph on the opposite page shows a crab carrying three anemones, it could barely move under the weight. This anemone can occasionally be found living on the claw of a spiny spider crab (pages 100-101). [Anemone up to 8 cm tall]

Parasitic anemone - *Calliactis parasitica*

Living alone

Three anemones on a single crab

Cloak anemone - *Adamsia carciniopados*

The anemone's white tentacles are visible between the hermit crab's legs

This species, common around most of Britain, also lives on hermit crabs, but the association is much more intimate than that between the parasitic anemone and its host. The cloak anemone's host is the small hermit crab, *Pagurus prideaux* (page 109). The anemone is never found living without a crab, and tends to stay with the same crab for life. Its base, which is usually brown or white with garish magenta spots, is wrapped right around the crab's mollusc shell home, hence the name "cloak". The anemone's white tentacles are positioned down between the legs of the crab in an ideal position to pick up the scraps which inevitably result from the crab's feeding activities. By secreting a hard extension to the crab's residence, the anemone removes the crab's usual need to seek larger shells as it grows. This reduces stress for the crab and of course also prevents the anemone itself from being abandoned. When touched, the anemone releases sticky white threads from its "cloak" as a defence mechanism. This will often be seen to occur if a host hermit bumps into a rock or worm tube as it retreats hurriedly across the seabed (top photograph, opposite page). On some anemones, the bright pink spots are obscured and it is only the emergence of these threads that give away their presence. The bottom photograph on the opposite page shows that a cloak anemone does not make its host invincible. A harbour crab (pages 92-93) is tucking into the remains of a small hermit crab while its cloak anemone appears to have been carefully peeled off and discarded. The anemone has discharged some threads in a last act of defiance. ["Cloak" up to 7 cm across]

Cloak anemone - *Adamsia carciniopados*

Defences activated

Not everyone is deterred. This harbour crab has removed and discarded the anemone, before tucking into the hermit crab

Sea loch anemone - *Protanthea simplex*

As its name suggests, this sea anemone is known mainly from Scottish sea lochs where it can be extremely abundant, particularly on vertical rock faces. Colour is pale orange through to white, with the column usually a denser shade than the long and slender translucent tentacles which cannot be retracted. Food particles that bump into any part of

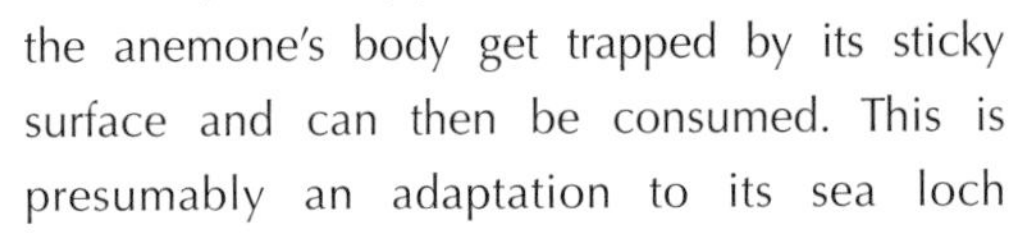

the anemone's body get trapped by its sticky surface and can then be consumed. This is presumably an adaptation to its sea loch environment where suspended food may be less abundant than on open coasts. [Up to 7 cm across, usually smaller]

Fireworks anemone - *Pachycerianthus multiplicatus*

Anemone with slender sea pens

The most striking feature of this spectacular species is its huge size. It is a burrowing anemone (like *Cerianthus lloydii*, opposite) and lives on muddy seabeds within its own tube. The long outer tentacles, all white or white with brown bands, may coil up if disturbed but cannot be retracted into the column. The shorter and stiffer inner tentacles are usually pale brown. It is quite a rare sight, mainly restricted to fairly deep water (10 to 130 m) in Scottish sea lochs. [Tentacle crown up to 30 cm across, column up to 30 cm tall, tube up to 1 m long!]

Burrowing anemone - *Cerianthus lloydii*

Dislodged anemone and its tube

This very common species of burrowing anemone (like the unusual fireworks anemone, opposite below) is not classified as a true sea anemone. Rather than attaching itself to a rock or similar firm surface, it lives in a soft felt-like tube that is constructed by specially adapted discharge capsules. Although the tube can reach up to 40 cm in length, only its uppermost rim will protrude above the sand or mud in which the remainder lies buried. The anemone's tentacles are usually all that is visible. The innermost set are short and stiff, while the outer ones are longer and sometimes attractively banded. The tentacles themselves are not retractable but, when disturbed, the whole anemone shoots back into its tube. The smaller photograph, of a dead anemone in its dislodged tube, shows the relative proportions of anemone and tube. [Tentacle crown up to 10 cm across, body/column up to 15 cm long]

Jewel anemone - *Corynactis viridis*

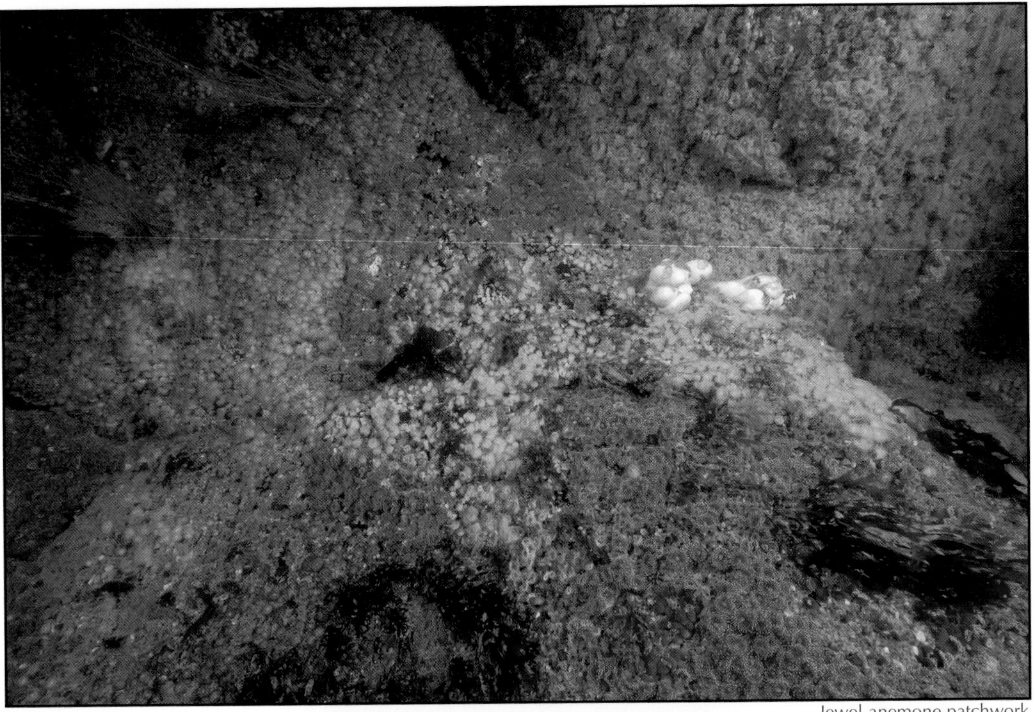
Jewel anemone patchwork

Small individually but impressive collectively, jewel anemones produce some of the most beautiful underwater scenery around Britain. The anemones occur in an amazing variety of colours: pinks, purples, reds, oranges, yellows, greens, browns and more can all be seen on a single rock face. Their capacity for prolific budding (asexual reproduction) means that the different colours are not totally mixed but occur in distinct patches. When this effect is combined with very clear water, the result is astounding, though they can also be found on the shore and even in quite muddy locations as long as there is good water movement. Jewel anemones are most impressive in locations off the south west of England, Wales and western coasts of Scotland, and are largely absent from the east coast of Britain. An individual jewel anemone is an attractive animal in its own right too. It has one hundred or so translucent tentacles, with distinctive white or brightly coloured rounded knobs on the tips; these knobs are laden with large numbers of discharge capsules of different types. The mouth is located on a minute cone in the centre of the tentacles. Jewel anemones are actually more closely related to corals than to the other anemones. [Up to 2.5 cm across]

Jewel anemone - *Corynactis viridis*

White cluster anemone - *Parazoanthus anguicomus*

This species belongs to a group of anemone-like creatures (sometimes called colonial or encrusting anemones) that are not true sea anemones. Clusters of its attractive white polyps arise from a shared encrusting base and a careful peek amongst the stems of a colony reveals that they are joined. The tentacles are in two rings with one set tending to reach straight up while the other set is more splayed. This species is generally found in quite deep water and this photograph was taken on a Cornish wreck in 35 metres where the colony covered large areas of superstructure. However, I have often seen it in shallower water off the west coast of Scotland. [Up to 3 cm tall]

Yellow cluster anemone - *Parazoanthus axinellae*

Another attractive colonial anemone, this species is usually pale yellow or yellowish-orange with darker colouration around the mouth. It is very similar to its close relative, the white cluster anemone (opposite) and, confusingly, can sometimes also be white. The shape then helps distinguish it, with the yellow cluster anemone being smaller, slimmer and having relatively longer tentacles than the white cluster anemone. The two photographs here are of the same colony in south Devon, the smaller one showing its typical location: under a slight overhang near the base of a rock face, with encrusting sponges and soft corals nearby. [Up to 2 cm tall]

Devonshire cup-coral - *Caryophyllia smithii*

Abundant around much of Britain and not only Devon, this species is the only common stony coral in our waters. Stony corals are similar to sea anemones but produce hard chalky skeletons to support and protect their bodies. They are usually colonial animals, and their fused skeletons can then form large coral structures. Cup-corals, however, live alone and do not fuse together, although many can live within close proximity. Their hard skeleton consists almost entirely of calcium carbonate and is cup or goblet-shaped, with pronounced ridges radiating out from the centre and running over the rim. As the creature lays down more skeletal material at its base, it is pushed upward and stays in the top of the cup. With tentacles fully extended, cup corals resemble small anemones. There are many different colour forms and some have a very attractive zigzag pattern of dense colour around the central mouth, while others are all white. The tentacles end in small but obvious knobs. The underlying skeleton may be largely obscured or quite obvious (bottom photo). Empty cup skeletons can sometimes be seen after the polyp has died. [Up to 4 cm across]

Sunset cup-coral - *Leptopsammia pruvoti*

With its suitably attractive name, this fantastically exotic-looking animal is very rare and numbers may even be decreasing. Found in just a handful of locations in the south west of the British Isles, it is a Biodiversity Action Plan listed species. The sunset cup-coral is larger than the Devonshire cup-coral (opposite) and is brilliant yellow or yellow-orange, sometimes with slightly darker orange trim around the mouth and rim. It lives on rock faces in fairly deep water, often around 30 to 40 metres, and there may be a good number of individuals grouped together even though it is not a colonial species. It is thought they may live for a very long time, over 100 years, but breed only occasionally so populations can be very vulnerable. The scarlet and gold star-coral (*Balanophyllia regia*) is also yellow but has a darker disc and is smaller with fewer tentacles; this is usually found in shallow water where there is strong surge. [Sunset cup-coral up to 5 cm across. Scarlet and gold star-coral up to 2.5 cm across]

Dead men's fingers - *Alcyonium digitatum*

Like some species of anemone, the soft coral colonies known as dead men's fingers can cover large expanses of rocky cliffs or seabed and so create their own brand of underwater scenery. Many of the rock faces around St Abbs in south east Scotland (see photograph opposite) are excellent examples. Dead men's fingers are generally orange or white. In northern Britain, both of these colour forms are very common but the white form seems more abundant in the south. Each "finger" is a colony of tiny animals that has formed a mutual skeleton of gelatinous material (hence *soft* coral) strengthened by embedded calcareous spicules. Until a colony reaches a height of about 5 cm it remains unbranched but, once larger than this, it tends to divide into several lobes. If arranged in a single plane as they often are, a group of lobes then appears like a hand. When active and feeding, the animals that make up the colonies extend their translucent tentacles and the lobes have a characteristically attractive "furry" appearance. In the photograph (above) the animals in the lobe at the lower right corner have retracted their tentacles. In the autumn, most colonies stop feeding and withdraw their tentacles for several months, while they prepare to spawn. Entire rock faces can then be covered by what look like knobs of expanded polystyrene. When feeding is resumed, an outer skin is shed, along with any encrusting growths that settled while the tentacles were out of action. As with so many animals that rely on suspended food, they are more abundant in areas of moving water. [Up to 20 cm tall]

Dead men's fingers - *Alcyonium digitatum*

White and orange dead men's fingers

Red fingers - *Alcyonium glomeratum*

Red fingers cover this rock face, with white dead men's fingers at the corners

Close-up of tentacles

Red fingers - *Alcyonium glomeratum*

Colonies showing tentacles extended and withdrawn

Not to be confused with the orange form of dead men's fingers, red fingers are a separate but closely related species of soft coral. The feeding tentacles are white, as opposed to the translucence of those belonging to dead men's fingers, and make a striking contrast against their red background. This allows extreme close-ups (opposite bottom) to show their full beauty. Further differences between the two species are that red finger colonies can be taller, and often appear slimmer, than those of their close relative and that they have a distinctly knobbly appearance when the tentacles are withdrawn (see bottom lobes in small photograph above). Red fingers are far less common than dead men's fingers and are only found on the western side of Britain. [Up to 30 cm tall]

Pink sea fingers - *Alcyonium hibernicum*

This small species of soft coral is rare and has only been recorded in a few places in the far west of the British Isles. It is coloured a delicate pink, with a thin white rim sometimes visible below the little spread of tentacles on each polyp. This photograph was taken in a set of sea caves in the Isle of Man where the pink sea fingers are unusually abundant. [Up to 4 cm tall]

Pink sea fan - *Eunicella verrucosa*

An array of pink sea fans of different sizes, near Plymouth

Rocky slopes dotted with pink sea fans are the sort of wonderfully exotic sight that is usually associated with warmer seas. Around Britain, this species is restricted to south west England and south Wales. It is most common at depths greater than 10 metres, although the odd individual is found in shallow water. Like dead men's fingers, sea fans are colonies of tiny creatures. The sea fan is classified as a gorgonian or horny coral. A skeleton composed of a dark brown protein (gorgonin), reinforced with calcium carbonate, runs through the fan and is covered with fleshy tissue from which the tentacles of the numerous polyps emerge to feed. This tissue gives the fan its colour; pink, pale orange or occasionally white (see top photograph opposite). Fans usually only branch in one plane which is at right angles to the prevailing current, thus giving each animal the maximum opportunity for feeding. The fans grow slowly, at around 1 cm per year, and the largest may be over 100 years old. They can easily be broken or dislodged from the rock by careless fin or arm movements and will die once knocked

Pink sea fan - *Eunicella verrucosa*

flat, so please take care! The pink sea fan is one of the very few marine animals protected by the Wildlife and Countryside Act, and it has also been identified as a priority Biodiversity Action Plan species. Sea fans are often used by dogfish as an anchoring point for their "mermaid's purse" egg cases. In some locations, a tiny species of sea slug lives on the fans and feed on their polyps, although these are thought to re-grow so the fan survives. The slugs' colouration and shape match those of the polyps so they are very difficult to spot. [Fans up to 50 cm tall in most areas but up to nearly 1 m tall in the Channel Islands]

Pink and white *Eunicella verrucosa*

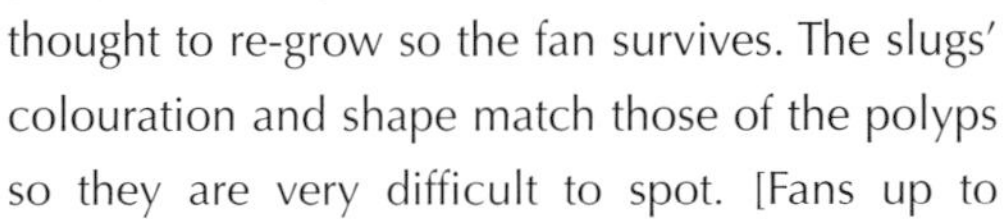

Northern sea fan - *Swiftia pallida*

Always white or pale grey, the northern sea fan looks like a pale and rather straggly version of the pink sea fan. It is widely distributed around Europe as a whole but, on British coasts, it is only common on the west coast of Scotland where it is found in sheltered and fairly deep water. By contrast, the pink sea fan is only common in the south west of Britain so you are unlikely to see the two species together. There is a small area of overlap in south west Ireland. The northern sea fan does not have the same legal protection as its pink relative but is just as vulnerable to physical damage and pollution. [Northern sea fans up to 20 cm tall]

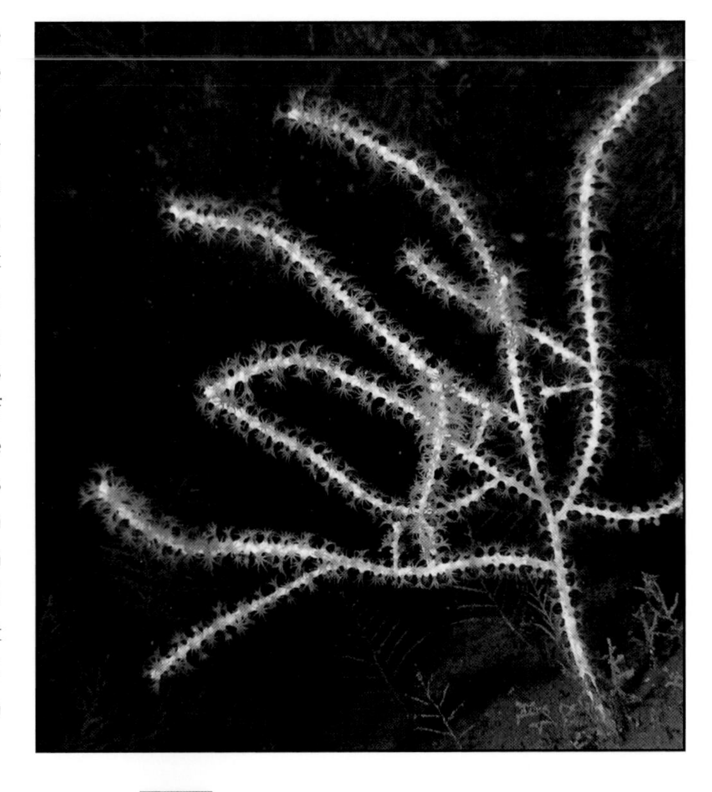

Slender sea pen - *Virgularia mirabilis*

The sea pens are feather-shaped colonies of anemone-like animals and are related to the soft corals and sea fans. Unlike those animals, sea pens are adapted for life on a muddy or sandy seabed. The lower part of the stem acts as an anchor buried in the sediment while the remainder bears the polyps, usually on side-branches. The slender sea pen has short side-branches which produce the slim feathery appearance. It can withdraw completely into the mud when disturbed. Several of these sea pens are often found living quite close together, as in the photograph. [Up to 50 cm tall]

Tall sea pen - *Funiculina quadrangularis*

This superb creature is quite an unusual finding but is memorable and spectacular. Reaching up to over 2 metres tall, its stem has a slight curve as it leans gently in any current. Its huge size should distinguish it from the slender sea pen, and it also has a much less fluffy appearance. The square cross section of its stem (that gives rise to the second part of its scientific name) is a further distinctive feature. In the background of the photograph, a few slender sea pens can be seen, dwarfed by their neighbour. The tall sea pen has been recorded from as far away as Japan and New Zealand but, around Britain, is largely restricted to very sheltered water on the west coast of Scotland. It is thought to be very vunerable to trawling activity. [Up to over 2 metres tall]

Phosphorescent sea pen - *Pennatula phosphorea*

The thick stem and large side branches of this species give it a much more bulky appearance than that of the slender and tall sea pens. Its bright white polyps contrast attractively with the deep red "body". As its name suggests, it can produce brilliant flashes and pulses of light if touched at night. Common in Scottish sea lochs and sheltered inlets, it can be found around most of Britain but is apparently absent from the south coast. [Up to 30 cm tall]

Oaten pipe hydroid - *Tubularia indivisa*

This species is often found in large aggregations on rock faces exposed to strong currents. Numerous long and very thin straw-like stems rise up from a mat of tangled fibres to form dense bunches. These stems support polyps that have pink bodies bearing a crown of long white, rather droopy, tentacles. The reproductive parts appear like a bunch of grapes near the centre of the crown. This species lives for about a year and produces a creeping larva rather than a free-swimming medusa. A number of types of sea slug prey on this hydroid and can often be found amongst the stems, or crawling up them to feed on the polyps (page 171). The slugs are not deterred by the hydroid's stinging cells, some even incorporating them for their own defensive use. [Up to 15 cm tall]

Solitary stalked hydroid - *Corymorpha nutans*

All the other bottom-living hydroid stages described in this section are colonial. This hydroid is unusual in being solitary during its sedentary polyp stage. *Corymorpha nutans* has a delicate appearance, resembling a tall and very slender sea anemone, with its tentacle-bearing top often drooping downwards. There are attractive pin-stripes running up its tapering column, which is found anchored in silt, sand or gravel. [Up to 10 cm tall]

Hermit crab fir - *Hydractinia echinata*

Almost always found on the borrowed mollusc shells of hermit crabs, this hydroid species forms an obvious and distinctive pale pink or white "fur". A colony consists of an encrusting mat which supports different types of polyp. Some are tall with a tiny crown of anemone-like tentacles while others are short and tightly coiled. The various polyp types predominate on different parts of the shell and the presence of the crab and its breathing currents are thought to affect their distribution and development. The larvae of this hydroid can detect a suitable hermit shell by its movement and attach themselves with stinging threads. They then form a colony by budding. [Tallest polyps reach up to 1.5 cm]

Kelp fir - *Obelia geniculata*

The obvious hairy "fuzz" often found on fronds of kelp and other brown seaweed is produced by colonies of this hydroid species. The "fuzz" is a forest of tiny stems with numerous small side branches that each bear a tiny polyp. The stems have a very characteristic zigzag form which distinguishes this species from other similar hydroids. A dusting of silt can sometimes collect on the colonies and this unfortunately obscures their wonderful structure. Unlike the hydroids above and on the previous page, this species belongs to the type (thecate) whose polyps have tiny protective cups into which they can withdraw. Distribution is worldwide. [Stems up to 5 cm tall]

Antenna hydroid (or sea beard) - *Nemertesia antennina*

This hydroid is found growing in distinct clumps. The main stems, pale orange or buff in colour, are unbranched but have numerous feathery side branches. Each stem in the clump is a colony of individual polyps, whose tentacles add to the furry appearance. The clumps provide shelter and food for small animals of all kinds and the white beads of sea slug eggs can often be seen entwined round individual stems. When looking at this species, it is very easy to see why hydroids are also known as sea-firs. The brightly coloured creatures, visible on the rock face around the sea beard clumps, are jewel anemones (pages 60-61). [Up to 25 cm tall]

Branched antenna hydroid - *Nemertesia ramosa*

This species is closely related and very similar to the antenna hydroid, but it can easily be distinguished because the main stems of its colonies are branched. This branching and the further sub-branching is quite irregular, creating an untidy appearance. Like the sea beard, it can be found attached to rock or to stones and shells in sand. It is also preyed upon by a variety of sea slug species. The colony in this photograph is being browsed by a nudibranch slug, *Eubranchus tricolor* (see also page 173). [Colonies up to 15 cm tall]

Hydroid medusae

In addition to "true" jellyfish, such as those shown on pages 78-82, and the comb jellies described on page 83, the jelly-like medusae of hydroids can also be seen drifting in the sea. These medusae (known as hydromedusae) are the free-swimming stages of hydroids that also have a bottom-living polyp stage at a different time in their life cycle. All the hydroids shown on pages 74-76 are at the bottom-living polyp stage (some species have no free-swimming medusa stage at all). The photographs on this page show three examples of hydroid medusae: *Aequorea forskalea*, *Staurophora mertensii* (sometimes known as the "white-cross jellyfish") and the smaller *Neoturris pileata*. Sometimes, very little is known about the bottom-living polyp stage of quite common hydroid medusae. Even the polyps' identity may be uncertain, and it is simply suspected that a particular hydroid polyp (which may be called by another name) is the bottom-living stage of a particular medusa. Hydroid medusae are typically simpler, smaller, more transparent and delicate in appearance than the "true" jellyfish; but swim and capture their prey in the same way. The medusa is the sexually reproductive stage of the hydroid. There are usually separate sexes whose eggs and sperm produce larvae which settle on the seabed to form new colonies of bottom-living hydroid polyps. The polyps then give rise to medusae by asexual budding, and the life cycle is complete.

Aequorea forskalea, 15 cm across

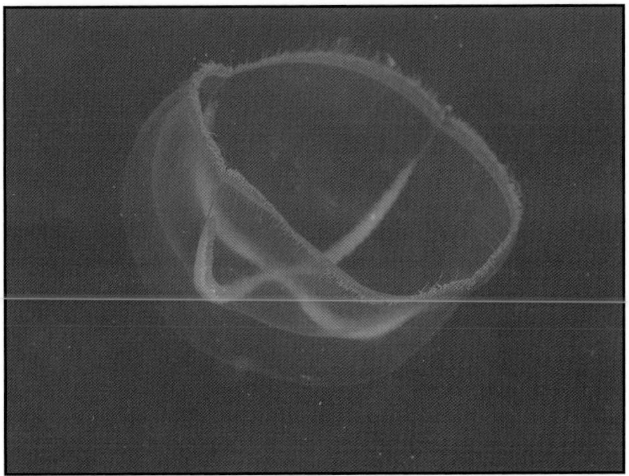

Staurophora mertensii, 10 cm across

Neoturris pileata, bell 3 cm long

Moon (or common) jellyfish - *Aurelia aurita*

This is the most commonly seen of our jellyfish and it can be very abundant in the summer, sometimes in huge swarms (see bottom photograph). Without long trailing tentacles, the moon jelly's most distinctive features are its reproductive organs. These appear as four relatively opaque horseshoe-shaped tissues, visible in its almost transparent umbrella. The umbrella itself is saucer-shaped and has numerous small tentacles around the edge, like a fringe. There are also eight sense organs around the rim and these are marked by slight indentations. The four arms near the central mouth are used in feeding, though planktonic food can also be captured in mucus anywhere on the umbrella and be transported to the mouth by a system of grooves lined with tiny beating hair-like cilia. The short peripheral tentacles can also capture prey with their stinging cells, though these are not of sufficient power to bother humans. [Up to 25 cm across]

Compass jellyfish - *Chrysaora hysoscella*

This species is easily identified by the attractive radial pattern of dark brown V-shaped markings on its "umbrella" or "bell". An additional dark circle in the centre of the pattern completes the appearance of an old-fashioned compass rose. Twenty-four slender (marginal) tentacles hang down from the edge of the umbrella while there are four much more noticeable (oral) arms in the centre. The marginal tentacles extend when the animal is hungry and, on capturing prey which may be any sort of planktonic animal, they contract in order to pass the food to the oral arms. When feeding is finished, the marginal tentacles remain contracted. Eight sense organs, each one situated between groups of three marginal tentacles around the umbrella's fringe, enable the jellyfish to maintain its orientation in the water. Young fish belonging to the cod family, particularly whiting, can often be found swimming around and between the compass jellyfish's tentacles (see bottom photograph). This provides the youngsters with protection from predators such as larger fish, who will avoid the tentacles, but it seems to be uncertain as to how the young fish themselves avoid being stung and eaten by the jellyfish. The sting of this species can sometimes cause a painful reaction on exposed human skin. [Umbrella up to 30 cm across]

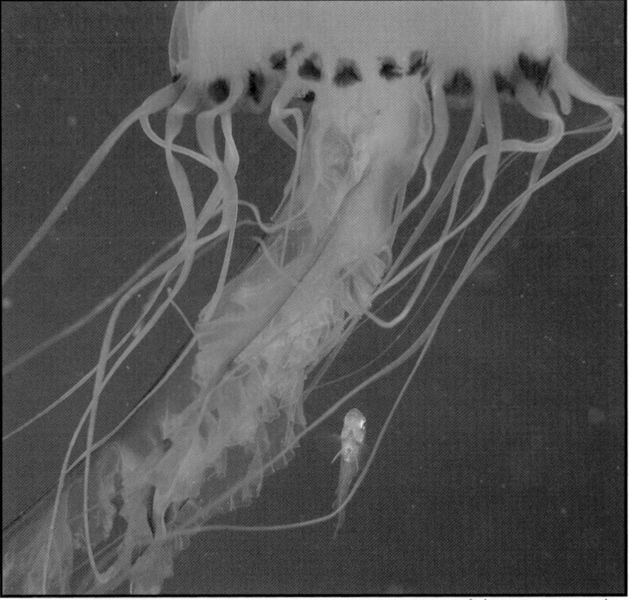

Young fish among tentacles

Mauve stinger - *Pelagia noctiluca*

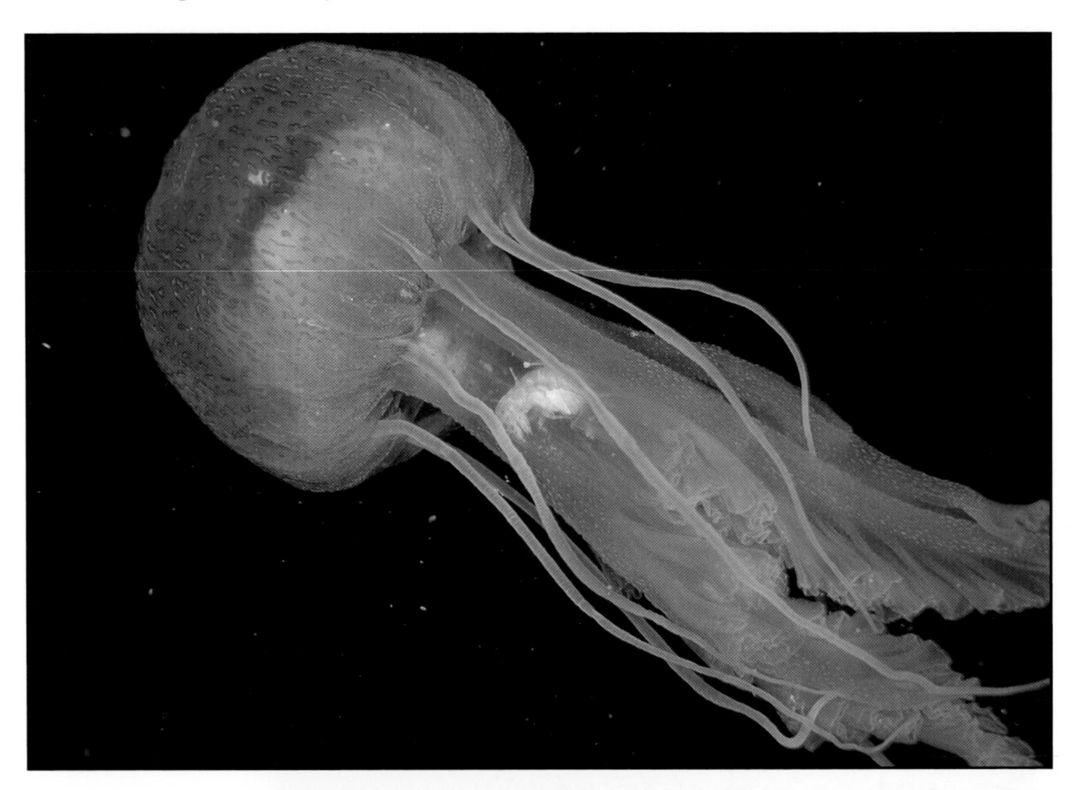

Although smaller than the other jellyfish described here, this species has a wonderfully striking appearance. The umbrella or bell is shaped like a globe (as opposed to the bowl-shape of many species) and is covered with warts which contain concentrations of stinging cells. The overall colour may be quite purplish, hence the common name. Its sting is powerful and can produce a very severe reaction. This does not seem to trouble the small shrimp-like crustaceans (amphipods) that live in association with this species (see top photograph). As the second part of its scientific name suggests, it can glow brightly at night if disturbed. [Up to 10 cm across]

Lion's mane - *Cyanea capillata*

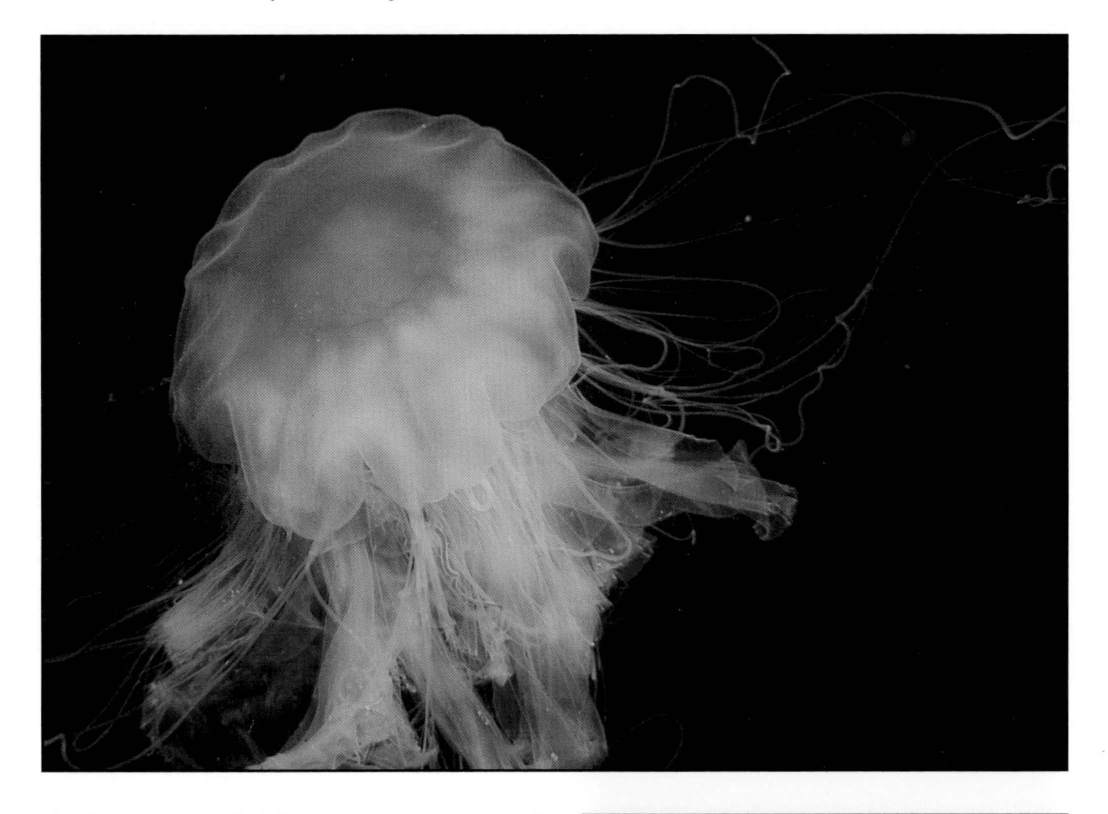

The lion's mane jellyfish is an impressive animal but it can leave a painful impression if approached too closely! Its tentacles, covered with powerful stinging cells, stretch up to three metres long when extended and it is quite easy to swim into them before the main body of the jellyfish is spotted. Fragments of tentacles, left on buoy ropes for example, also retain their stinging power. The lion's mane umbrella usually has brown markings and is rather flat with its edge formed into large lobes. There are four large arms surrounding the mouth, but these are far shorter than the tentacles. The tentacles are arranged in eight bunches, with each bunch containing over a hundred tentacles, the oldest of which is often coloured dark red. As with the compass jellyfish (page 79), small fish are often seen sheltering amongst the tentacles (see bottom photograph). The lion's mane is quite common off north, east and west coasts of Britain but is rare in the Channel. [Up to 50 cm across]

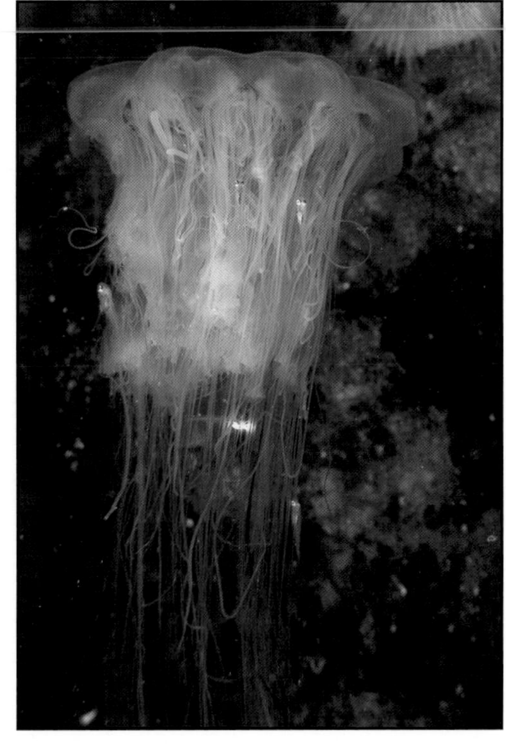

Blue lion's mane - *Cyanea lamarckii*

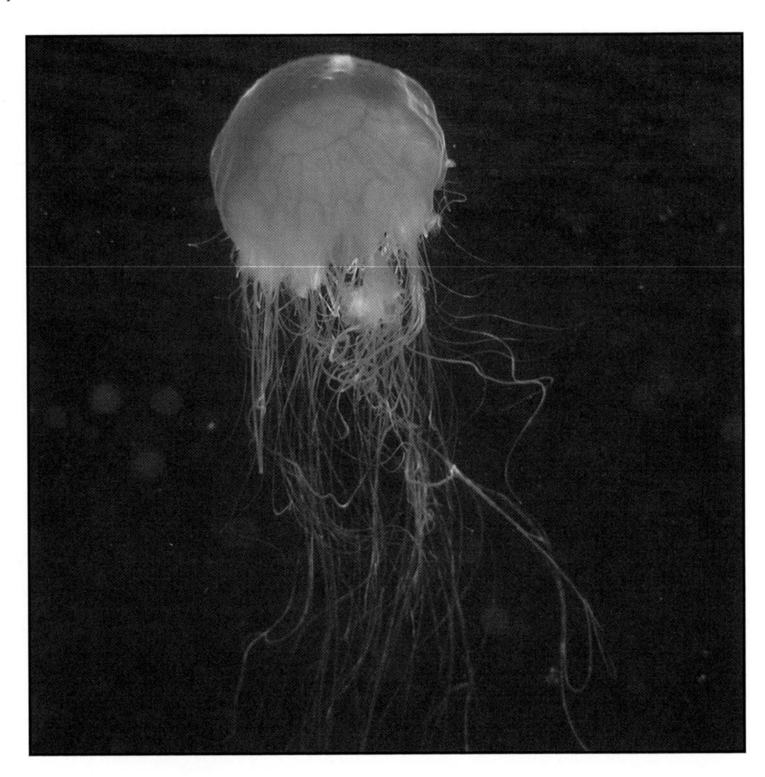

The umbrella of this animal, sometimes known as the "blue jellyfish", is usually blue or purplish but can also be yellow. It is a close relative of the lion's mane and is very similar in form although considerably smaller. There are fewer (about 50) tentacles in each of its bunches and their sting is much weaker than the lion's mane's. It is found all around Britain and tends to be more common than the lion's mane in the south. [Up to 30, but usually less than 20, cm across]

Barrel jellyfish - *Rhizostoma octopus*

This species is the largest jellyfish found in British waters. Not often found close to the shore, it is usually seen from boats or by divers decompressing on a shot-line, for whom its massive pulsating form makes an interesting diversion. There are no markings on the whitish umbrella except for its dark rim, and no peripheral tentacles. The eight central arms are fused for much of their length to form a dense bunched mass. This structure bears hundreds of tiny mouth openings, each surrounded by miniature tentacles bearing stinging cells. Creatures striking the tentacles and small enough to pass through a mouth opening will be ingested. The sting is harmless to humans and there are even reports of these jellyfish being eaten by eighteenth century fishermen. *Rhizostoma* is apparently very sensitive to the vibration from ships and will move downwards as one approaches. [Up to 80 cm across]

Sea gooseberry - *Pleurobrachia pileus* (a comb jelly)

The sea gooseberry is a small round comb jelly. Strictly speaking, comb jellies should not be in this chapter at all. They belong to the phylum Ctenophora, and not to the phylum Cnidaria which contains all the sea anemones, corals, hydroids and true jellyfish. Nevertheless, comb jellies resemble small jellyfish and their group is related to the cnidarians. The characteristic feature of comb jellies is the set of (usually 8) comb rows that run down the outside of the gelatinous body. Each row is a series of small plates formed from the fusion of tiny hair-like cilia. The plates beat rhythmically to drive the animal mouth-first through open-water, with some comb rows beating faster than others if a turn is required. The shimmering combs are often beautifully iridescent in sunlight and those of some species give off luminescence at night if disturbed. Many comb jellies have tentacles that, while lacking the stinging cells of cnidarians, have adhesive cells for catching their prey which may even include small fish. *Pleurobrachia pileus* can occur in large numbers from spring through to autumn. It has a pair of branched tentacles that reach up to 50 cm long and form an effective net for trapping prey, mainly tiny planktonic crustaceans. These tentacles can also be completely retracted into pouches. Sea gooseberries are eaten by jellyfish such as the moon jellyfish (page 78) and some fish. As well as just referring to this species, the name "sea gooseberry" is occasionally used as an alternative to "comb jelly" in describing the whole group. [Body up to 3 cm long]

Bolinopsis infundibulum (further example of a comb jelly)

This much larger comb jelly is sometimes abundant near the surface during the spring and summer. The photograph shows how the body can appear almost completely transparent, and a hint of iridescence on the combs is just visible. [Up to 15 cm long]

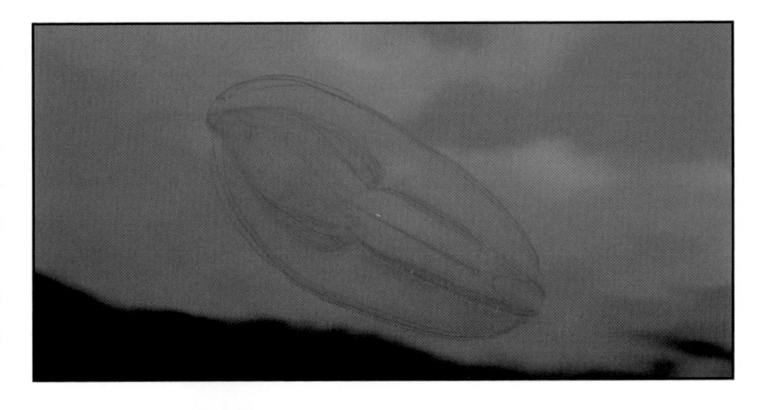

Chapter 4
CRUSTACEANS
Crabs, lobsters, prawns, shrimps & barnacles

Crustaceans can be thought of as the aquatic animals which wear suits of armour. This armour is jointed, and crustaceans belong to the larger grouping, Arthropoda, which means "jointed leg" (insects are also arthropods and are now thought to have originated from within the crustaceans). All the familiar large mobile animals such as lobsters, crawfish, crabs, prawns and shrimps belong to the same sub-group of the crustaceans, the Decapoda, and have a similar body plan. They have ten legs (hence the name), the first pair of which are often claws. The head and thorax are fused and covered by a single section of armour, known as the carapace. The abdomen, or tail, is protected by further sections of armour and may be obvious as in the lobsters, prawns and shrimps or much reduced and tucked up as a flap underneath the rest of the body, as in the crabs. Barnacles are in a separate sub-group from these more familiar members and are described in more detail on page 122. While they are an initially surprising inclusion in the crustacean category, a close view of their feeding limbs (see top photograph opposite) provides a good clue to their identity. In addition to the decapods (crabs, lobsters, shrimps etc.) and the barnacles, there are a huge number of other crustacean species which are tiny animals that live in the plankton or on the seabed and make up vital parts of the marine food web.

A suit of armour, the benefits and drawbacks

The suit of armour worn by crustaceans has obvious benefits, providing good protection against predators, rigid anchorage points for powerful muscles and hard surfaces for crushing, cutting and grinding their prey. The major drawback is that it has to be shed periodically to allow growth. As a crustacean grows to fill its shell, it forms a soft leathery coat beneath the outer casing. At moulting time, the shell splits at a pre-determined point and the animal, clad in its soft coat, eases itself out. The middle photograph, opposite, shows a prawn found struggling to free itself from its old armour. All the limbs, tiny mouth parts and even the eyes have to be removed from their casing. The bottom photograph, opposite, shows the discarded armour suit of a harbour crab underwater; the opening at the back where the owner climbed out is visible. Such a suit is so complete that it can easily be mistaken for a dead crab. Having left the old suit, a soft crustacean swells itself up with water to create some growing room (see middle photograph page 86) and the process of hardening up new armour then begins, taking several days in the case of large crabs and lobsters. The soft animal is obviously very vulnerable and has to stay as well hidden as possible. Once hard, it will have a particularly large appetite and seek food vigorously before eventually becoming relatively inactive and listless as it prepares for the next moult. Reproduction is affected by moulting, because the female of many species is only receptive after she has just shed her shell, so the moulting cycles of crustaceans can really be seen to dominate their lives. The top photograph on page 86 shows the sequence of armour suits worn and discarded by a shore crab kept in an aquarium for three years. The amount of growth between successive moults is substantial.

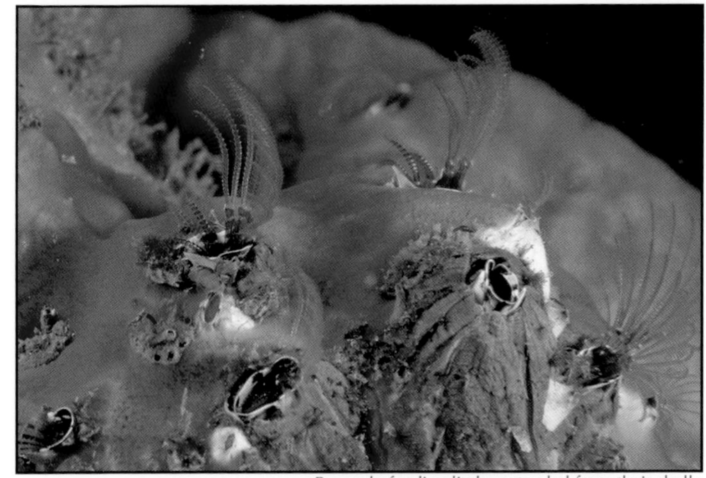

Barnacle feeding limbs extended from their shells (which are encrusted by orange sponge)

Moulting prawn struggling free from its armour

Regenerating limbs

Although moulting inserts dangerous interludes into the life of a crustacean, it also provides the opportunity for regenerating limbs. A leg or claw that is damaged or seized by a predator can be jettisoned by the animal, which breaks it off deliberately at a particular point near to its base. A new limb can then start to form

The suit of armour discarded by a moulting harbour crab

Series of discarded armour suits showing the growth of a shore crab

inside the shell and will emerge at the next moult. This is why crustaceans will often be seen with a very small leg or claw, though this will catch up with the other limbs over subsequent moults. The process can lead to a "right-handed" crab or lobster becoming "left-handed". If the large crushing claw is lost, the return to a fully equipped state can be hastened by converting the smaller cutting claw to a large crusher and then growing a new small cutting claw on the other side. The aquarium-kept crab whose moult series is shown above did exactly this after losing his right-hand claw to a fish (see 5th armour suit then onwards).

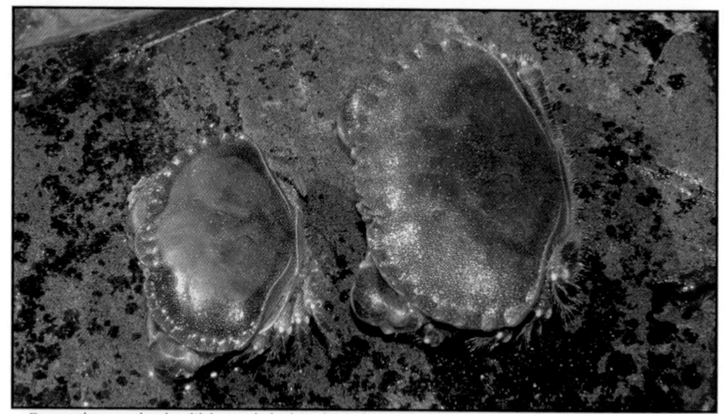

Recently moulted edible crab behind its discarded suit of armour, found on the shore

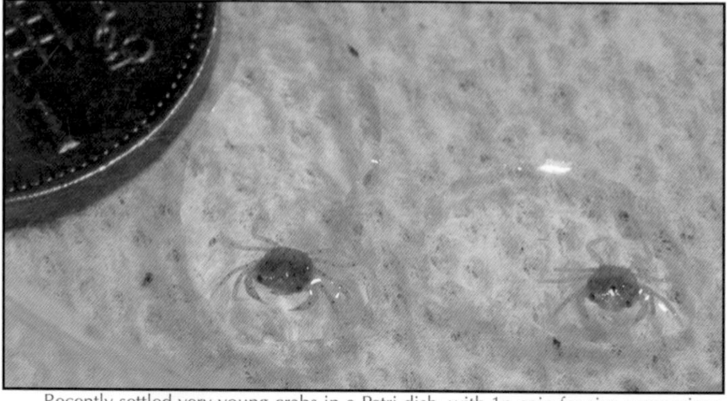

Recently settled very young crabs in a Petri dish, with 1p coin for size comparison

Breeding

All the crustaceans included in this chapter have a larval phase that drifts as plankton in the open sea. On first hatching from the eggs, these larvae look very different from their parents. They develop through various larval stages, feeding, growing and moulting at each one. Gradually taking on more adult features, they eventually settle down on the seabed as immature miniature versions of the adult. After many more moult cycles, they are ready to breed. The bottom photograph opposite shows very small young crabs, and the drawing below shows a typical crab life cycle (the number of larval stages and adult moult cycles varies from species to species). Velvet swimming crabs in pre-mating embrace, mating and carrying eggs are shown on page 91; and shore crabs (page 88), harbour crabs (pages 92-93) and hermit crabs (page 105) are shown with their partners before mating.

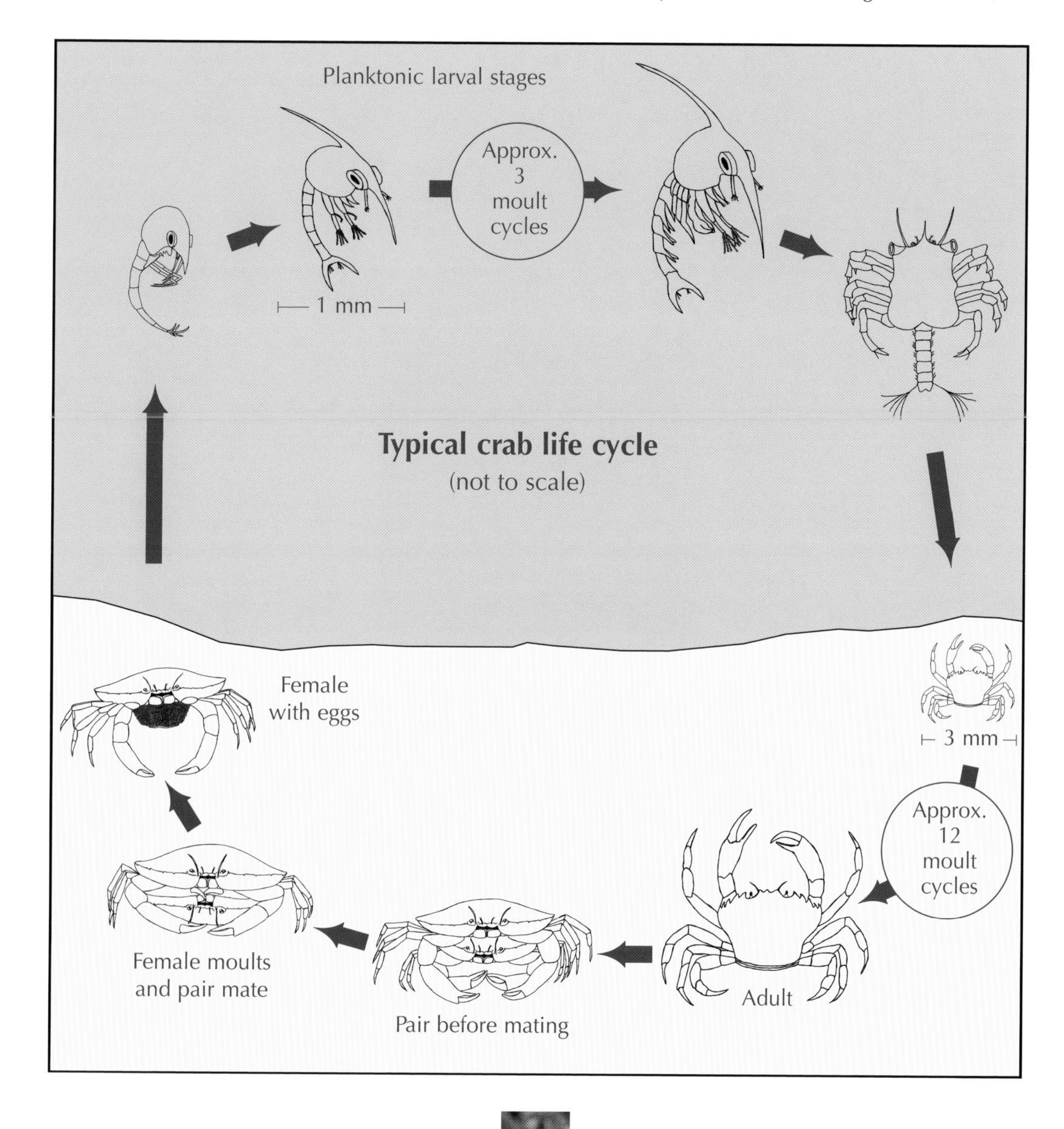

Shore crab - *Carcinus maenas*

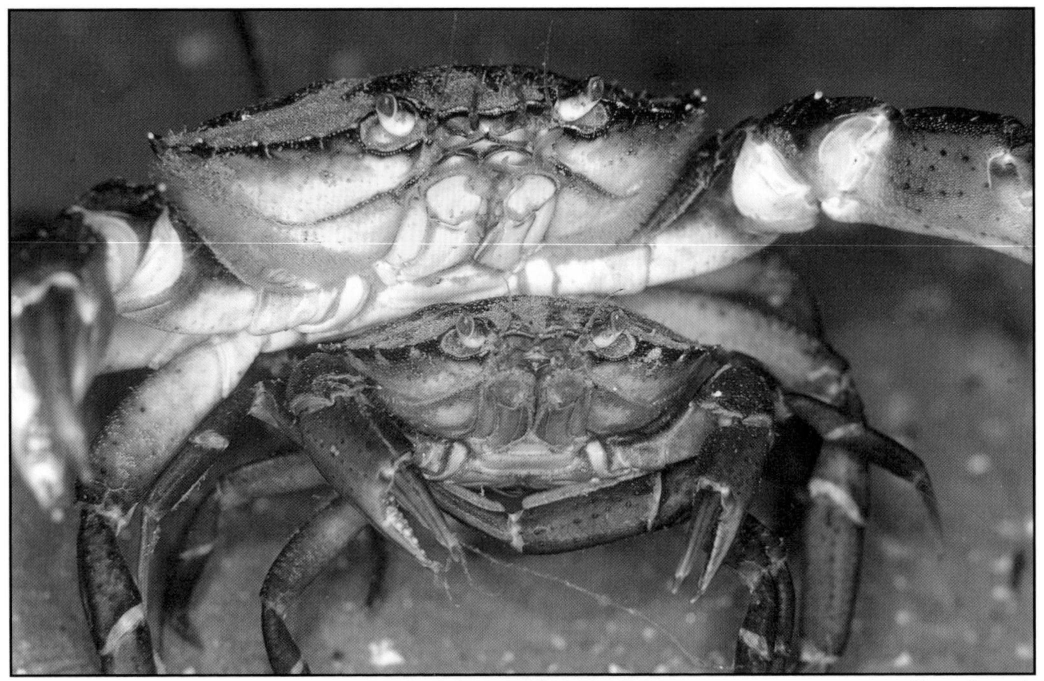

Pre-mating pair

The shore crab is a familiar animal to anyone who has ever looked under stones on a beach or dangled a bacon-baited string into a muddy creek. Very common on the shore and in shallow water, it is seen less often by divers than many crabs because it is most abundant in muddy estuaries and inlets. The pre-mating pair in the main photograph show how colour can vary amongst shore crabs. The larger male has the typical green and yellow colouration while the female underneath has the deeper orange worn by some crabs that are thought to not have moulted for a considerable time. The male is holding onto the female until she moults and then they can mate. The shore crab's carapace shape can be seen clearly in the photograph of discarded moults on page 86. Though it belongs to the swimming crab group, the paddles on this crab's rearmost legs are less well developed than those of the velvet swimming crab and harbour crab (pages 89-93) and it is a much poorer swimmer. Shore crabs are extremely hardy and, by tolerating a wide range of conditions in terms of salinity and exposure to air, can live in virtually fresh water or barely any water at all. They eat a variety of animal prey such as worms, shrimps and any molluscs that they can break open. Cannibalism also occurs; the crab in the smaller photograph probably tore the claw he was eating from a colleague who was soft after moulting. Some populations of shore crabs suffer a high level of infestation from the parasitic barnacle, *Sacculina carcini* (page 123). [Carapace up to 8 cm across]

Small shore crab eating a colleague's claw

Pennant's swimming crab - *Portumnus latipes*

Left: discarded armour showing distinctive carapace shape. Right: crab underwater.

This small crab is often found by children digging in the sand at low tide. At first glance, it can be mistaken for a small pale shore crab (to which it is related), but the elongated heart-shaped carapace is distinctive. As can be seen in the photograph of discarded armour, the carapace is about as long as it is broad (unlike the shore crab's which is markedly broader than long). The Pennant's swimming crab's back usually has an attractive pale spotty pattern which gives good camouflage on a sandy seabed. It is unusual to see these crabs underwater and they will quickly dig back into the sand if disturbed. [Carapace up to 2 cm across]

Velvet swimming crab - *Necora puber*

Velvet swimming crab eating a small shore crab

The body of this abundant and widespread crab is covered with short greyish-brown hairs that provide the velvet-like appearance. The most **notable features** however, are the bright red eyes

Velvet swimming crab - *Necora puber*

and the blue lines on its legs and claws. As in all swimming crabs, the final section of the rear-most pair of legs is flattened to form a **swimming paddle**. Velvet swimming crabs tend to be pugnacious and, when encountered, may rear up and spread their claws in defiance rather than shrinking away into a crevice. Very common on rocky and stony seabeds, they can also be found on sandy and muddy bottoms where they may dig themselves in. They are **versatile feeders** and, being fast-moving and agile, can catch prey such as fish, prawns and other crabs. Depending partly on its size, a preyed upon crab may be captured, killed and eaten; or merely lose a few limbs to the velvet swimmer before managing to struggle free (see photographs on page 89 and top of this page). All sorts of bottom-living animals including worms and molluscs are also eaten. The middle photograph on this page shows a velvet swimmer tackling a top-shell. Lacking the strength to simply crush such a shell, the crab gradually chips away at its opening with one claw while the other claw is used for holding and turning. Although often actively carnivorous, some populations of velvet swimmers have been found to

This velvet swimming crab was eating the claw and leg it had just pulled off a shore crab

Crab breaking into top-shell

Crab eating red seaweed

Velvet swimming crab - *Necora puber*

eat large quantities of seaweed (see bottom photograph opposite). Pairs of crabs in **pre-mating embrace** (top photograph, this page) are a common sight. As in many crab species, the female can only mate when soft just after shedding her armour. A male will therefore find a female approaching a moult and carry her tucked beneath his body, moving around much as normal. He will help her out of the old armour when the time arrives and then, after mating, protect her while her new armour starts to harden. The middle photograph shows a pair of velvet swimmers actually mating. This is easily distinguished from the pre-mating embrace because the female is upside down and therefore lying underside-to-underside with the male. The bottom photograph shows the ultimate result of such a union, a female is carrying a large mass of **fertilised eggs** (noticeably granular and usually orange) between her abdomen and underside. An egg-carrying female is said to be "in berry". [Carapace up to 10 cm across]

Pair before mating

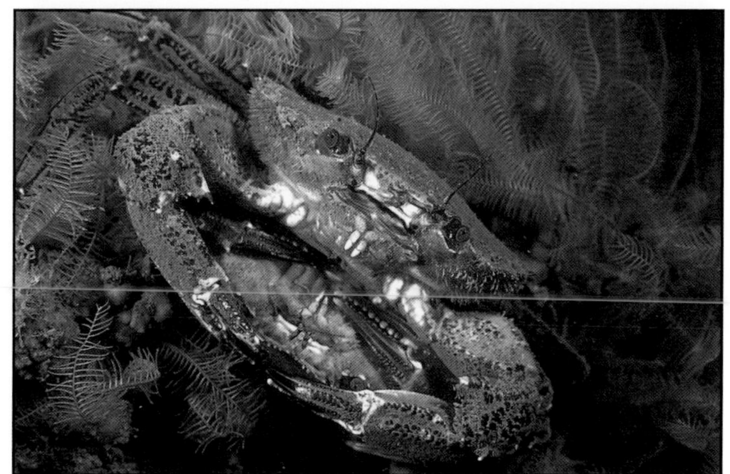

Pair in the midst of mating

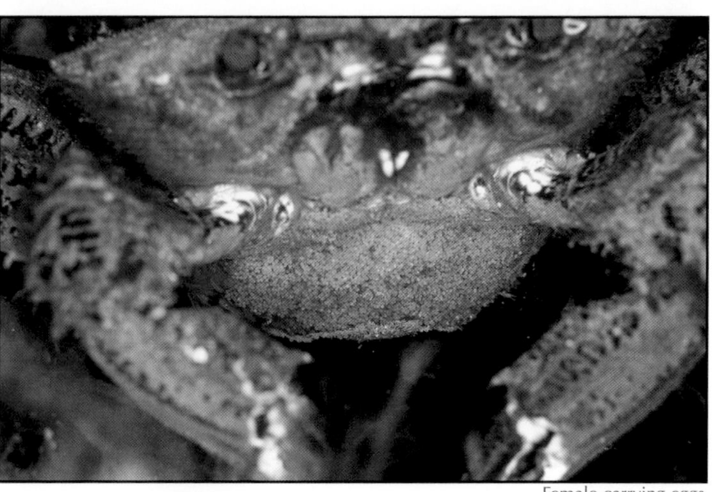

Female carrying eggs

Harbour crab - *Liocarcinus depurator*

Pre-mating pair in defensive posture; the female is joining in even though she has only one claw

This species is similar in shape to the velvet swimming crab but is more lightly built and pale pink or reddish brown in colour. These swimming crabs are almost always found on sandy seabeds, sometimes buried with only antennae and pale brown eyes showing. If disturbed they have various options: making a rapid escape, assuming an aggressive posture with claws spread wide or simply moving a short distance and submerging into the sediment. They are very capable swimmers with efficient paddles on their rear-most pair of legs. With a distinctive swimming style the body is propelled rapidly sideways, just above the seabed, with the other legs held stiffly out from the body for maximum streamlining (see middle photograph this page). When swimming, the paddles are a blur of motion but, when stationary, the characteristic mauve spots on the blades can be seen (see bottom photograph this page). Like the velvet swimmer, pairs of these crabs are often seen in pre-mating embrace (top two photographs this page, and opposite). [Carapace up to 7 cm across]

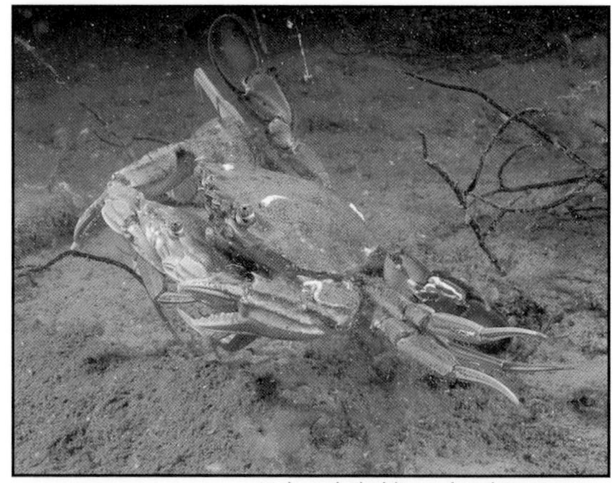

Male crab (holding a female) swimming

The distinctive purple spots on the swimming paddles are obvious here

Harbour crab - *Liocarcinus depurator*

Pre-mating pair of crabs, being followed by another male who seemed as though he was waiting for an opportunity to "steal" the female

Edible crab - *Cancer pagurus*

The edible crab has a thick, oval shell with a very distinctive "pie-crust" edging. The claws are large but the legs appear relatively small. The overall colour is a pink-brown while the claws have noticeable black tips. Compared to the swimming crabs, edible crabs are built very much for strength rather than speed. This is reflected in their prey; molluscs that might require hours of patient manipulation from a swimming crab can be quickly crushed by an edible crab's powerful claws. Large mussels are often eaten (see opposite top) and even very thick shells, such as that of the whelk (see bottom photograph this page), may be broken open with more effort. Conversely, they would find chasing small fish much more of a challenge

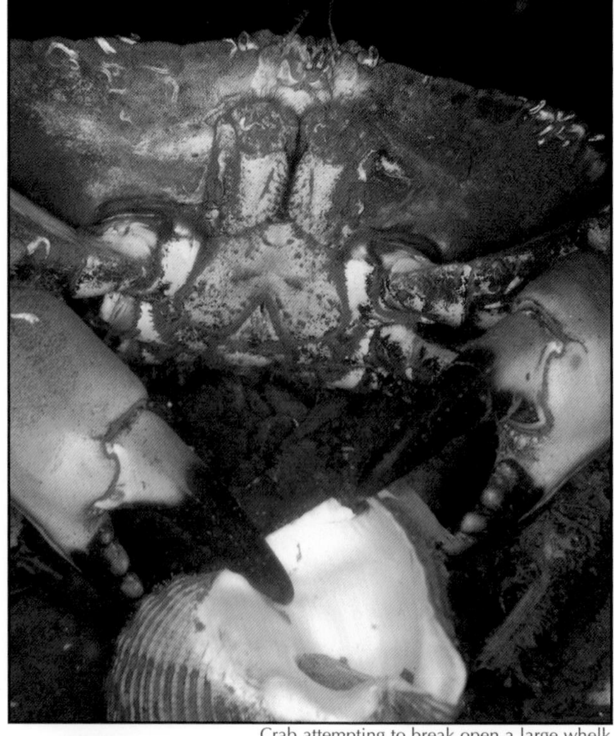

Crab attempting to break open a large whelk

Edible crab - *Cancer pagurus*

than their nimbler relatives. Edible crabs are found in a variety of habitats at all depths. On rocky bottoms, they live in crevices while, on soft seabeds, they dig themselves into the sediment. Here, they may also dig huge pits in their hunt for burrowing prey such as clams and worms. A muddy seabed can then become a mass of craters, each containing a busily excavating edible crab (see middle photograph this page). As in the swimming crabs, edible crabs mate just after the female has moulted, with the male guarding the female before she moults and after they have mated. This guarding often seems to involve him wedging his mate in the back of a hole or crevice (bottom photograph this page) rather than carrying her around as a male swimming crab does. [Carapace up to 25 cm across, but rarely more than 15 cm]

Crab eating mussels

Digging pit in muddy seabed in search of food

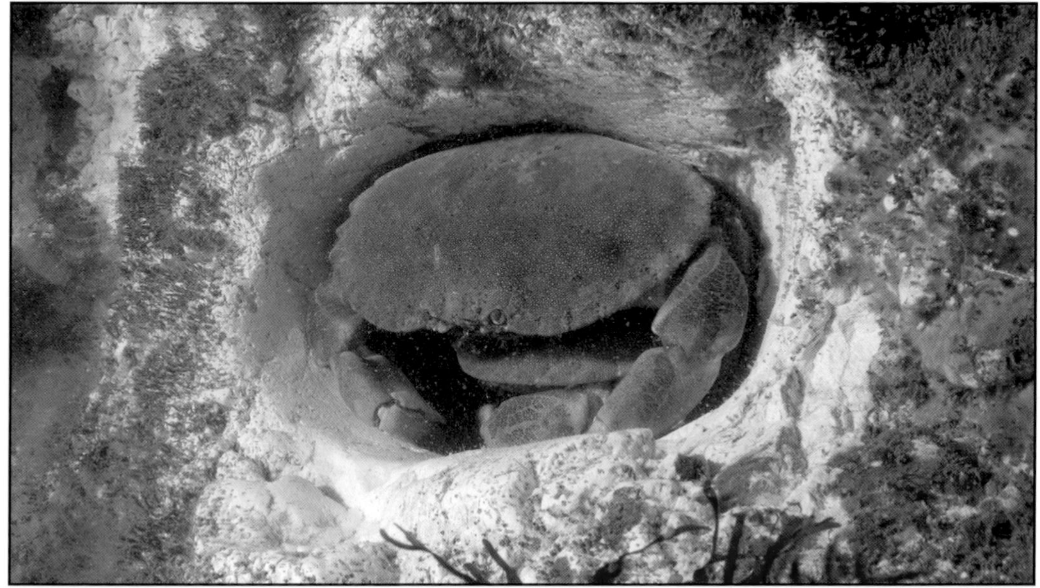
Pair of crabs in a hole on a chalk reef

Masked crab - *Corystes cassivelaunus*

Male carrying female around mating time

Two males fighting

Masked crab - *Corystes cassivelaunus*

Female

A male crab tries to dig down...

Masked crabs can be common on sandy seabeds but spend much of their time buried and hidden from view. Their distinctively proportioned carapace is longer than it is broad and so, when tilted upright, is the perfect shape to allow the crab to slip rapidly into the sand. Short legs give them a relatively slow lumbering gait but work well for digging. While buried, their long antennae are held together to form a snorkel-like tube, through which the crab can draw clean water for respiration; it is usually just the very tips of the antennae that protrude from the sand. The male has extremely long claws, much longer than the female's. Unlike many crab species, female masked crabs do not have to be soft shelled (just moulted) in order to mate. They are still only receptive at certain times, however, and a male will carry a female about either grasped in a claw (see top photograph opposite) or tucked underneath his body as in other crabs. Masked crabs are often only spotted when inadvertently disturbed from the sand, and they may just move a short distance before digging in again. On occasion, they attempt to dig directly into the hiding place of another masked crab and the sequence of photographs on the right shows the dispute that resulted from such an intrusion; the bottom photograph opposite shows another fight between two males. The "masked" crab name comes from ridges on the carapace which, to the imaginative, look like a human face. [Carapace up to 5 cm long]

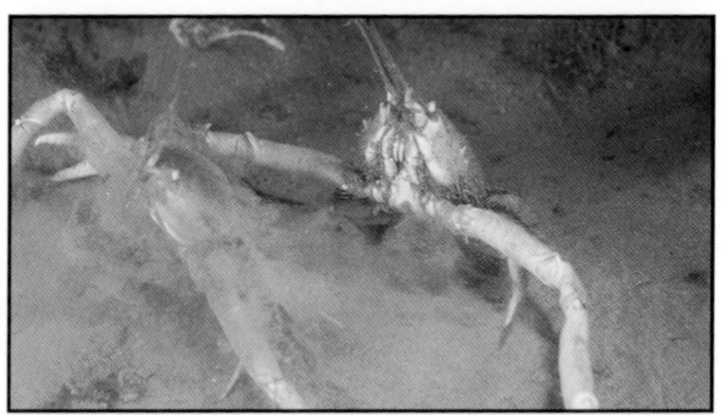
...into the hiding place of another...

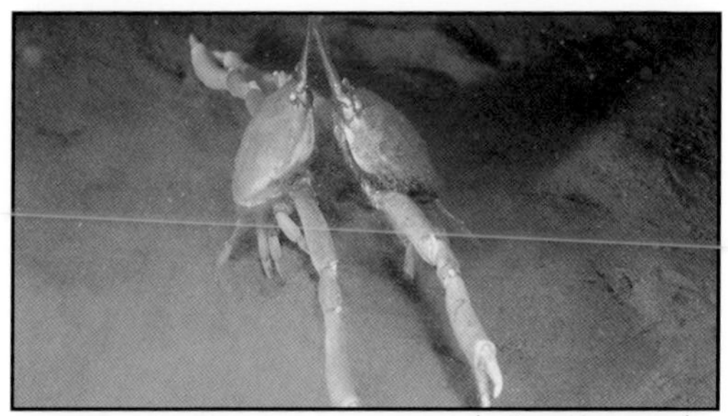
...who is not impressed...

...and a dispute results

Hairy (or bristly) crab - *Pilumnus hirtellus*

This small crab gets its name from the prominent bristles on its carapace, claws and legs. Silt tends to get trapped around these bristles and can help make the crab more difficult to see against its background. It is reddish-brown in colour and there are 5 sharp "teeth" on the front edge of its carapace either side of the eyes. The claws are quite stout and one (often the right, but here the left) is much larger than the other; although this can be true of crabs of any species which are re-growing a claw lost in a fight. Hairy crabs are often seen on rocky or stony shores but also live in deeper water. Their varied diet can include carrion, small mussels or other animals and plant material. [Carapace up to 3 cm across]

Hairy crab uncovered on the shore

Montagu's crab - *Xantho incisus*

Risso's crab - *Xantho pilipes*

Montagu's crab

Risso's crab

Crabs uncovered on the shore (both photographs)

Montagu's crab is sturdily built with robust, usually black-tipped claws and relatively short legs. The obvious grooves in its carapace (it is sometimes called the furrowed crab) add to the muscular appearance. Risso's crab (occasionally known as the "less furrowed crab") is similar but has less obvious grooves on its carapace, brown-tipped claws and is often yellowish with marbling rather than an even brown. The best distinguishing feature is the fringes of long hairs on the legs of Risso's crab; the legs of Montagu's crab bear just a few bristles. Both crabs are most often found underneath rocks or stones on the lower shore (with Risso's crab also reported from sandy seabeds). At first glance, they could be confused with very small edible crabs but they lack the edible's distinctive pie-crust shell edging. [Carapace up to 7 cm across, both species]

Square crab (angular crab or mud runner) - *Goneplax rhomboides*

The square crab has a most distinctive appearance and is unmistakable whether viewed from the front or the back. Its carapace is much broader than it is long and has very pointed corners, while the long eye-stalks and very long slender claws enhance the exotic image. Both photographs show male crabs. The female, which is seen less frequently, has much smaller claws. This species is quite an unusual sighting but it is abundant in some areas with muddy sand. While many of the crab species described in this chapter can dig themselves into soft sediments, the square crab is different in that it makes a permanent burrow. This burrow is normally a shallow 'U' shape with two exits, but there may also be side tunnels. Crabs are often seen sitting at the entrance to their burrows, but they can also be found hiding in crevices beneath boulders or out and about on patches of mud between the boulders. Their detached claws are regularly seen lying on the seabed in areas where they live, presumably lost in fights or during struggles with predators. [Carapace up to 5 cm across]

Spiny spider crab - *Maja squinado*

The carapace of this large crab is quite circular, shaped like a dome and covered in many short spines, with longer and more pointed spines around its edges. There are two large points between the eyes, looking almost like horns. It has long and slender legs, and the claws are not much bigger than the walking legs, except in large males (as above). Spiny spider crabs can be found on most seabed types, often resting in the open or wandering free rather than hiding in crevices. Despite these habits and their large size, some individuals can be very difficult to spot because of the camouflaging seaweed or sponge placed on their carapace and legs, which blends in perfectly with their surroundings (see inset to top photograph opposite). This camouflage is of course desirable, and a spider crab does not welcome the attempts of a rock cook wrasse to clean it by nibbling at the encrusting growths (see main top photo opposite). Large ballan wrasse are the usual, and more welcoming, recipients of the rock cook's cleaning activities (see pages 280-281). Spiny spider crabs can sometimes be seen grouped together in large heaps, or moving across the seabed in substantial numbers, presumably on their way to or from such a gathering. The heaps, which are found in the summer, may stay in the same place for a few months, often containing up to a hundred individuals. Huge mounds of up to a thousand crabs have even been reported. Crabs moult in the centre of the heaps and, while soft, are protected from predators by all the crabs outside them. The heaps also mean that newly moulted (and therefore receptive) females will have suitors readily available, so heaping seems to play an important part in the breeding of these crabs. The bottom photograph opposite shows part of an annual gathering near Brixham. Spiny spider crabs have a varied diet, feeding on seaweed and encrusting animals such as hydroids. They can also often be found eating carrion. [Carapace up to 20 cm across]

Spiny spider crab - *Maja squinado*

Rock cook wrasse attempting to "clean" a spider crab. Inset: crab with sponge camouflage

Part of a spider crab gathering; the water was murky because the crabs stirred up mud from the seabed as they moved

Sea toad (or great spider crab) - *Hyas araneus*

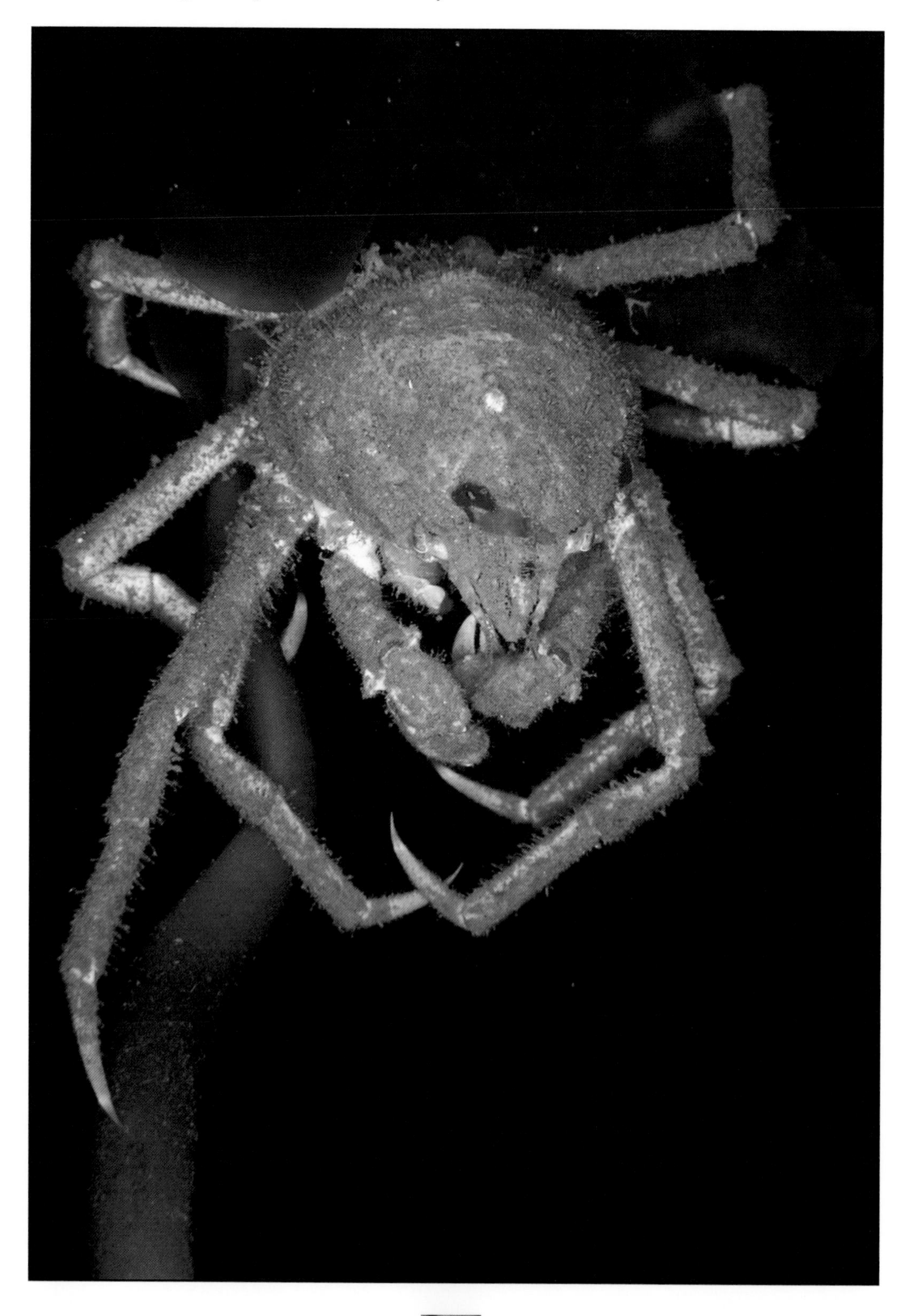

Sea toad (or great spider crab) - *Hyas araneus*

This crab is intermediate in form between the large bulky spiny spider crab (page 100) and the very spindly small spider crabs (below and page 104); so the "great" in one of its names is slightly misleading. The carapace, which may be disguised by seaweed, sponge or other growths, is much wider at the rear than at the front so is roughly triangular. These crabs are often seen clinging to kelp (see opposite) where they graze the small attached seaweeds as part of their omnivorous diet. The elevated position may also help them to catch passing jellyfish which they appear to enjoy eating. However, it is not clear whether they can catch healthy jellyfish or simply rely on coming across sickly individuals. The sea toad seems to be much more common off northern coasts than in the south. [Carapace up to 8 cm across]

Sea toad eating a jellyfish

Long-legged spider crab - *Macropodia rostrata*

The body of this "spindly-looking" spider crab is very small, especially in comparison to leg length, and its eye stalks are prominent. Although often very common, it can be difficult to spot due to having emaciated features and the habit of attaching seaweed or sponge to its carapace and legs as camouflage. The crab in this photograph is obvious because, having just moulted, it had not yet camouflaged its new armour. There are several very similar species of *Macropodia*; definite identification of this one was only possible from close examination of its discarded armour. [Carapace up to 1.5 cm across]

Recently moulted crab with its discarded armour (beside urchin)

Leach's spider crab - *Inachus phalangium*

Territorial dispute

This small spider crab has a triangular carapace. The legs are very long and slender but the claws, which are usually held folded under the body, are fairly sturdy, particularly in the male. Carapace, claws and even legs are often covered with sponge and pieces of algae, which may make the crab virtually invisible. These crabs are very often seen hiding around the base of a snakelocks anemone or, if it is a large anemone, sitting within its tentacles (main photograph this page and page 42). In some areas, virtually every snakelocks has its own resident crab. The two crabs in the smaller photograph on this page appeared to be in dispute over the favoured space beneath the anemone's tentacles. The nature of the relationship between crab and anemone seems to be something of a mystery. *Inachus phalangium* is very difficult to distinguish from *Inachus dorsettensis*, the scorpion spider crab, particularly as the distinguishing features on the carapace are usually obscured by camouflaging sponge. For the sharp-eyed, *I. dorsettensis* has a broad U-shaped tip to the promontory of its carapace between the eyes, while that of *I. phalangium* comes to a narrower point. [Carapace up to 3 cm across]

Common hermit crab - *Pagurus bernhardus*

A large male hermit crab dragging round a female prior to mating

Common hermit crab - *Pagurus bernhardus*

Two hermit crabs scavenging on a dead shore crab

The common hermit crab can be abundant on both rocky and sandy seabeds, and small individuals are often found in pools on the shore. Colour is typically pale orange-brown with darker reddish markings. The claws are covered with numerous bumps and lines of spines. Only the front part of a hermit crab is protected by the usual crustacean armour which is shed to allow growth. The top photograph opposite shows a crab with its recently shed armour in front of it. The hermit's soft rear end is protected by a disused mollusc shell into which the whole body can withdraw when danger looms; and the larger right-hand claw forms a barrier across the entrance. As a hermit crab grows, it seeks progressively larger shells to occupy. Young hermits live in small shells like those of winkles, while large adults inhabit whelk shells (see different shells occupied by pre-mating pair on page 105). Hermits only change home after a prolonged and thorough inspection of the new shell. The bottom photograph opposite shows a crab that seemed to be carrying out one of these inspections, an ambitious idea since the original mollusc owner was still in residence. The hermit's rear end is coiled so that it fits neatly into the shell and is equipped with small hooks for clinging to it, while the last two pairs of walking legs are used as struts to support the shell's weight. The top photograph on page 108 shows a hermit in such a badly damaged whelk shell (possibly from when a large crab ate the whelk) that its rear end is visible. It was found in a small inlet where there was a huge number of hermit crabs; intense demand for shells had presumably led to such a sub-standard home being accepted rather than rejected. The bottom photograph on page 108 shows a small hermit that appeared to be trying desperately to pull another from its shell. The struggle went on for over 20 minutes, but the crab being harassed always managed to wriggle free when the attacker got close to achieving its apparent goal. Hermit crabs are versatile feeders and can prey on other animals, scavenge on bottom deposits and carrion (see photograph above) or filter food from the seawater. Animals found living on or in the mollusc shells carried by the common hermit crab include the parasitic anemone (pages 54-55), the hermit crab fir (page 75) and the commensal ragworm (page 129). [Carapace up to 4 cm long but mollusc shell home makes crab look much larger]

Common hermit crab - *Pagurus bernhardus*

Recently-moulted hermit crab with discarded armour

Examining a potential home

Common hermit crab - *Pagurus bernhardus*

Hermit crab living in badly damaged shell, the usually hidden abdomen is visible

Small hermit crab appearing to try to pull another from its shell; with a third crab watching and waiting

Anemone hermit crab - *Pagurus prideaux*

The anemone hermit crab tends to be a darker reddish-brown than *Pagurus bernhardus*, the common hermit crab. The shape of its claws and carapace are often quoted as distinguishing features but I find the large dark eyes are the most obvious difference. What is more noticeable than the features of the crab itself, however, is that almost all individuals carry a cloak anemone (described in more detail on pages 56-57). The cloak anemone on the crab in the photograph is very obvious but this is not always so. With its white tentacles positioned down underneath the crab, the anemone can be very hard to spot, particularly if the distinctive magenta spots are obscured by a fine dusting of silt. The anemone stays with the hermit for life and frees it from the need of changing to larger second-hand mollusc shells as it grows (the armour on the front part of the crab still needs to be moulted of course). In return for providing the crab with additional security and growing space, the anemone gets extra food and these crabs have even been observed placing food into their cloak anemone's mouth. *Pagurus prideaux* is usually found on seabeds of muddy sand or mud below low tide level; small hermits seen in rock pools and on clean sand will very often be young *Pagurus bernhardus*. [Carapace up to 1.5 cm long but mollusc shell home makes crab look much larger]

Pagurus forbesii

This small hermit crab was found near the mouth of the Helford estuary in Cornwall. It is included as an example of an individual whose soft abdomen (tail) is protected by a growth of sponge, rather than the usual second-hand sea snail shell. The photograph also shows how its eggs (small dark spheres) are carried within the entrance to the shelter. Although its claw shape indicates the crab is *Pagurus forbesii*, this species does not seem to be recorded as inhabiting sponges. [Carapace up to 1 cm long]

Diogenes pugilator

Identification of this small species of hermit crab is easy because its left-hand claw is obviously larger than the right. Quite common in the south and west of Britain, it lives on sand and can bury itself extremely quickly when encountered. When I have come across these crabs, however, they have often been too busy fighting to take evasive action. The photograph above was taken of a small group gathered around a dead or dying netted dog-whelk. Despite there being more than enough food to go round, the crabs broke off from eating every few seconds to do battle! Their fighting skills and large left-hand claws apparently make these crabs successful in disputes over shell rights with other species of small hermits. [Carapace up to 1 cm long]

Broad-clawed porcelain crab - *Porcellana platycheles*

Broad-clawed porcelain crab uncovered on the shore

Porcelain crabs are more closely related to squat lobsters than they are to other crabs. Their last pair of legs is much reduced and folded away so they look as though they only have 3 pairs of walking legs. If handled roughly, they can readily shed their legs and this may have given rise to their name. This species is very small but its broad flattened claws fringed with hairs are extremely distinctive. The carapace is roughly circular and has hairs around its edge; the overall colour is greyish brown. Found underneath stones and boulders on rocky shores, and also where there is mud around the stones, the broad-clawed porcelain crab can be very abundant. It filter feeds on suspended particles and may also eat carrion. [Carapace up to 1.5 cm across but broad claws make it look larger]

Long-clawed porcelain crab - *Pisidia longicornis*

Long-clawed porcelain crab uncovered on the shore

This crab is even smaller than the broad-clawed porcelain crab and, without the broad claws to bulk it up, appears really tiny. The claws are long, slender and look quite delicate. The carapace is close to circular as in the broad-clawed but does not have the fringes of hairs and can look quite shiny. This photograph shows how porcelain crabs appear to have only 3 pairs of walking legs. The long-clawed porcelain crab is a suspension feeder. Often found under stones on the shore, this species' range also extends into deeper water although only extremely observant divers will spot them. They are reported as living among colonies of the potato crisp bryozoan (or Ross "coral"), see page 193. [Carapace up to 1 cm across]

Spiny squat lobster - *Galathea strigosa*

Squat lobsters look like a cross between crabs and lobsters. Their long body is flattened, with the tail tucked up underneath it (see smaller photograph). The tail can be used as a paddle for rapid backward escape. This widespread species is particularly striking, with a brilliant red carapace decorated by blue patches and stripes, and long red-tipped claws covered with a "fur" of brown spines. It is rarely seen in its full glory, however, because it usually hides away in narrow crevices, often clinging upside down to the rock ceiling and shrinking further out of view when approached. It is most likely to be seen out and about at night. [Body, including tail, up to 10 cm long]

Side-on view showing tail tucked up

Olive squat lobster - *Galathea squamifera*

This small squat lobster is very common on the shore and in shallow water, but it normally stays hidden from view and is only seen if stones are disturbed. It will then escape quickly, moving backwards with rapid flicks from its tail. The individual in the top photograph was unusually bold and those in the bottom photo were found hiding amongst *Serpula* worm tubes (see pages 136-137). Colour is usually a greenish brown but youngsters are more reddish. The base sections of the claws bear several large spikes along their inner edges that can be very obvious (see top photo). This species is thought to be mainly a filter feeder, eating suspended organic material. [Body, including tail, up to 6 cm long]

Long-clawed squat lobster - *Munida rugosa*

Dispute over a crevice

Long slender claws and bulbous eyes peering out from a crevice or hollow are the hallmarks of this abundant animal. The long-clawed squat lobster is an extremely common sight in areas where the seabed is a mixture of stony crevices and mud, such as around the inlets and sea lochs of Scotland's west coast. On muddy slopes with a scattering of boulders, an individual can be found in virtually every sheltering place. The overall colour is red or reddish brown but the claws, about double the length of the body, are the most distinctive feature. Its antennae are nearly as long as its claws. This species appears much less timid than the other squat lobsters and is content to remain partially out of its shelter if approached carefully, often spreading its claws in defiance. It may even edge forward to investigate the intrusion and is thus easy to photograph. Long-clawed squat lobsters seem to eat a variety of food including dead crabs and other carrion. Disputes between neighbours over food or the occupation of a hiding place can sometimes be observed (see above). [Body, including tail, up to 8 cm long but claws can make it look larger]

Long-clawed squat lobster - *Munida rugosa*

Close-up showing complex mouthparts typical of crustaceans

Eating part of a velvet swimming crab

Squat lobsters tucking into a moon jellyfish

Common lobster - *Homarus gammarus*

A "left-handed" lobster at home

A lobster out and about in shallow water

Common lobster - *Homarus gammarus*

A magnificent but all too rare sighting, the lobster's appearance is unmistakable: dark blue armour with pale yellow markings and long bright red antennae. The powerful claws are quite different in shape; one (usually the right-hand one) is heavier and is used for crushing, while the other is a sharper cutting tool. Lobsters are normally found hiding within rocky holes or inside wrecks during the day but, when watched for a while, may move cautiously forward to investigate the intruder. They will usually only emerge from their lair at night but can occasionally be seen roaming the seabed during the day (see bottom photograph, opposite page). Such encounters usually seem to happen when one lobster has been disturbed from its hiding place by another. They feed on a variety of bottom-living animals, alive or dead, and are also well known cannibals; it is difficult to keep more than one lobster in an aquarium because the first one to moult will usually be rapidly consumed by its companion. They generally move by walking, but can also swim rapidly backwards in short bursts by using the powerful tail as a paddle. Mating is thought to occur in the late summer but females can store the sperm over the winter, so the eggs are not fertilised until the following summer. They are then held externally and may take a further year to hatch into free-swimming larvae. After a few moults while in the plankton, these transform into miniature lobsters and settle on the seabed. Large lobsters moult only very occasionally, if at all, and can become covered with barnacles and other encrusting organisms. They may live for up to 50 years. [Body up to 75 cm long but rarely more than 30 cm]

Crawfish (or spiny lobster) - *Palinurus elephas*

The crawfish is a large animal with strong, spiky body armour but no claws. Colouration is orange brown with yellow markings. The antennae are longer than the body and heavily built, with large spines at their bases so that they can be used as defensive weapons. Crawfish live in rocky areas and may often be found in holes, but they are also more likely to be seen wandering free over the seabed during the day than the common lobster. They are rare in water shallower than 20 metres. Numbers of these animals have been greatly reduced, with intensive fishing an important contributing factor. Unlike the common lobster, the crawfish is only found on western coasts of Britain. [Body up to 50 cm long but rarely more than 35 cm]

Scampi (Norway lobster, Dublin Bay prawn) - *Nephrops norvegicus*

Scampi at entrance to its burrow

The host of common names used to describe this species is probably a reflection of its considerable commercial importance across much of Europe. It is also called langoustine or simply "nephrops". The body colour is orange and the large eyes are jet black. It is a relative of the common lobster and, though smaller and much more slimly built, is similar in form. The slender claws are about as long as the rest of the body and carry rows of distinctive spikes. Scampi live in deep water (usually greater than 30 metres) and in burrows on muddy bottoms. They are therefore rarely seen by divers. The burrows can form complex systems of horizontal tunnels and scampi will spend most of their time there. They come out to hunt, mostly at night, for food such as small crustaceans and worms. Fries's gobies (see page 301) commonly live in the scampi burrows. Scampi have suffered from overfishing in many areas, and trawling for them also damages or kills other inhabitants of the seabed. [Body up to 20 cm long]

Common prawn - *Palaemon serratus*

Prawn in crevice with lobster (both photographs)

These animals are a very common sight in the nooks and crannies of rocky reefs and wrecks, and are often found in groups. The almost transparent body has numerous delicate reddish brown lines and there may also be obvious yellow bands and blue markings on the legs. The front two pairs of legs bear nippers, which are used to pick up small pieces of food as the prawn walks across the seabed. Prawns can also swim, either backwards in rapid bursts using the tail fan, or forwards using the small flaps beneath the abdomen for propulsion. Perhaps surprisingly, groups of prawns often seem to occupy the same crevice as predators such as velvet swimming crabs, lobsters (both photographs this page) and conger eels (see page 247). The prawns probably benefit from the larger animals' leftovers. [Up to 10 cm long]

Humpback (pink or northern) prawn - *Pandalus montagui*

This attractive crustacean lives up to all these versions of its common name. It has a humped back and, although translucent, there are characteristic pink patches on its carapace and bright red lines marking its body. Channel coasts represent the southern extent of its distribution and it seems to be most commonly seen while diving in the north of Britain. Unlike the common prawn (page 119), it is rarely found in pools on the shore. It is thought that some pink prawns are female throughout their life while others start as males and become females later. They live for about 3 or 4 years, having bred from their first year. This species is also known as the Aesop prawn. [Up to 16 cm long, but usually about 5 cm]

Brown shrimp - *Crangon crangon*

Dug in, only the eyes are left showing

The brown or common shrimp is abundant in sandy areas but can be very difficult to spot because of its mottled sandy colouration and tendency to remain buried with only the eyes exposed. The shrimp in the main photograph was more obvious than usual because it was walking across dark sand that had been turned over by an excavating crab. The head and body of the brown shrimp are more flattened than those of a prawn and they lie almost flush with the seabed, rather than being lifted up by the legs. The front pair of legs bear claws that are more substantial than those carried by many small prawns. It has a varied diet which includes animals such as small worms and crustaceans plus algae and detritus. If a brown shrimp is seen walking across the sand it may rapidly dig itself in (as shown in the sequence of smaller photographs), using an odd shuffling motion to get started, followed by sweeping movements of its long antennae to brush sand over its back. [Up to 9 cm long]

Barnacle - *Balanus perforatus*

Barnacles belong to a totally different sub-group of crustaceans from all the other animals described in this chapter. On first glance, one would not even think of them as crustaceans at all. With their sedentary life-style and limpet-like shape, barnacles look rather like molluscs, which was how they were classified until 1830. As with other examples in the animal kingdom, it was an examination of their larvae (which look like those of other crustaceans) that revealed their true identity. A barnacle has been described as being "like a shrimp which glues its head to a rock, lives in a house and kicks food into its mouth". This is entirely accurate, as glands on the young barnacle's head produce a special cement for permanent attachment, the usual crustacean armour is modified to form flat plates that make up the "house", and the limbs form a sieve for catching plankton. When observed underwater the rhythmic sweeping of the feeding "sieve", looking like a grasping hand, can be seen (see photos above and on page 85). The limbs are rapidly withdrawn if sudden movement is detected and small fish can sometimes be seen trying to nip them off before this happens. Barnacles occur in large populations on the shore and in shallow water. They are hermaphrodites (simultaneously male and female) and mating usually occurs between neighbours. The resulting embryos develop in the body of the parent until they become free-swimming planktonic larvae, and these then develop further until they are ready to take on adult life. It is at this point that they glue themselves head-first to a rock and quickly assume the familiar adult form. There are several very similar common species of barnacles, including other *Balanus* species, which have an identical basic lifestyle. *Balanus perforatus* has simply been included as an example of its kind. [Up to 3 cm across]

Sacculina carcini (a parasitic barnacle)

Swimming crab with reproductive mass of *Sacculina* visible beneath

Parasitised shore crab with heavy encrusting growth

Though classified as a barnacle, this bizarre animal has a strange parasitic form totally different from the normal hard-cased barnacles that live attached to rocks. Just as barnacles were identified as crustaceans from examination of their larvae, *Sacculina* is recognised as a barnacle by its larval form. A female *Sacculina* larva attaches to the surface of a young crab and then grows into branching roots which penetrate the tissues of the unfortunate host. The usual host species is the shore crab, *Carcinus maenas* (page 88), but other members of the swimming crab family are also affected. The parasite eventually produces a yellow or pale brown reproductive mass which is visible as a large lump beneath the crab's abdomen (see top photograph). The lump is distinguishable from the crab's own egg mass because it is smooth rather than granular. Eggs produced by ovaries in the lump are fertilised by male *Sacculina* larvae which have no adult stage. A crab affected by the parasite is prevented from moulting so its carapace may become heavily encrusted with animals such as barnacles and tube worms (see bottom photograph). This is usually a more obvious sign of infestation than the lump of the parasite itself. In addition to interfering with moulting, *Sacculina* also feminises male crabs and they take on certain female characteristics such as a broader abdomen. *Sacculina carcini* is the best known parasitic barnacle but there are several other similar species which affect different groups of crabs, squat lobsters and prawns.

Chapter 5
WORMS

Few groups of animals would initially seem to be of less interest than the worms, but some of those that live in the sea are surprisingly attractive and definitely worthy of a closer look. The term "worm" is very misleading because it tends to be applied to any creature that is long and wriggly. Different groups of animals, which vary enormously in terms of their biology and level of sophistication, fall into this category as a result. For ease of reference, different types of worm are all included in this chapter. Representatives of the flatworms and ribbon worms are dealt with on page 126-127 and the rest of the chapter belongs to the segmented worms. The characteristics of the different groups are described here.

Flatworms

Flatworms form the phylum Platyhelminthes and are fairly simple, typically leaf-shaped animals. They are regarded as being more advanced than animals such as cnidarians (sea anemones, jellyfish) because their tissues work together to form organs that carry out particular functions such as digestion. However, because flatworms are either very small or greatly flattened, there is no need for the circulation system or body cavity possessed by many other animals such as segmented worms, crustaceans and molluscs etc. Flatworms like the candy stripe (page 126) are free-living but many species, such as tapeworms and liver flukes, are well known parasites of land animals.

Ribbon worms

Ribbon worms make up the phylum Nemertea and, as their name implies, are thin and long. They are reasonably simple but slightly more sophisticated than the flatworms. Unlike flatworms, which have a single opening to their digestive system, ribbon worms have a separate anus and mouth. This means the intestine can use a more efficient one-way system where undigested and digested food is separated. Ribbon worms also have a circulation system, and more advanced muscles and nerve networks than flatworms. A remarkable feature of ribbon worms is their proboscis, a trunk-like extending structure, which shoots out of the body under pressure to trap or stab their prey. Two ribbon worm species are described on page 127.

Segmented worms

Segmented worms, belonging to the phylum Annelida, are quite complex animals, more on a par with crustaceans and molluscs. Their body cavity permits the development of advanced internal organs, while the nervous system is capable of supporting a variety of behaviour patterns. Virtually all the segmented worms found in the sea, and all those species described in this

Organ pipe worm tentacles

chapter (except for the leech on page 139), belong to the bristle worm sub-group. Typically, bristle worms have a head which bears intricate jaws, antennae and sense organs, while the remainder of the body consists of a series of less specialised segments that are fairly similar to one another. Many of these worms move freely from place to place but, because they burrow into soft seabeds or stay hidden amongst the organic debris on rocky seabeds, they are rarely seen by snorkellers or divers. Some species leave a very obvious clue to their existence every year in the early spring. They attach their green egg capsules to seaweed (see smaller photograph) and these can be seen in huge numbers, forming an obvious part of the scenery. In general, however, the segmented worms that make themselves most obvious underwater are those that have given up the free-living way of life and live in a tube that they have constructed. The honeycomb worm (page 130) and all the bristle worms after it in the chapter are tube worms. Many of these species have heads and sense organs less well developed than those of their active relatives, but they have a crown of very specialised, often beautiful (see main photograph) feathery tentacles. These tentacles are extended out from the tube into the water, where they serve to both collect oxygen for respiration and catch suspended food particles.

Worm egg capsules attached to seaweed

Candy stripe flatworm - *Prostheceraeus vittatus*

Found on mud or amongst stones and seaweed, the candy stripe flatworm is so thin and flexible it gives the impression of almost flowing across the seabed. Colouration is distinctive, cream with very marked dark "pin-stripes", and its body is often thrown into folds at the edges. There are two tentacles at the head end of the body which make the worm look rather like a sea slug, but these are a totally different group of animals (see Chapter 6). Mobile flatworms such as *Prostheceraeus* are carnivorous and feed on slow-moving, sedentary or dead animals. Their smooth gliding motion is produced by thousands of tiny hair-like cilia on the underside of the body acting in unison. [Up to 5 cm long]

Bootlace worm - *Lineus longissimus*

Individuals of this very slender ribbon worm species can grow to over 30 metres in length, so it could be thought of as Britain's largest animal! More usually, they are around 5 metres long when extended, but the entire body is rarely visible and will contract rapidly if touched. Colouration is typically dark brown to almost black, sometimes with a slight iridescent sheen. The head is only slightly wider than the rest of the body and has several eyes on each side, though these are difficult to discern against the dark pigment. [Up to 30 m long]

Football jersey worm - *Tubulanus annulatus*

Although they may not instantly remind you of a football jersey, the markings on this ribbon worm are extremely distinctive. Three slender white stripes run along the reddish brown body, one on the back and one down each side. These are complemented by 50 or so well-spaced white rings that encircle its body. Apart from its markings, this species is quite similar in form to the more commonly encountered bootlace worm. The football jersey worm is sometimes found crawling across a sandy or stony seabed but is often hidden under stones or in crevices. It has a very wide distribution in the northern hemisphere and is found on the west coast of North America, all across the Atlantic and in the Mediterranean. [Up to 1 m long]

Sea mouse - *Aphrodita aculeata*

A strange-looking animal that does not even look like a worm. The upper side of its broad and flattened body is covered with a felt-like fur while, round its edges, are some stout dark bristles and a fringe of silky iridescent hairs that can appear to glow gold, yellow and blue. The telltale worm-like segments are only visible on the body's underside. The sea mouse lives on bottoms of muddy sand and is usually buried with only its hind end exposed; it is thought to prey on other worms as well as eating carrion. The array of body hairs prevents mud particles being drawn in with the animal's breathing current. [Up to 20 cm long]

Paddleworm - (Family Phyllodocidae)

Most of the other segmented worms described in this chapter live permanently in burrows or tubes. Paddleworms often hide under stones but they can also be seen roaming freely over the seabed. They are mainly active carnivores and can be watched inspecting hollows in the sand and making a grab for tiny shrimp-like crustaceans. These worms may produce large amounts of mucus when disturbed. The name paddleworm arises from the overlapping leaf-shaped paddles that fringe both sides of the body and can be used to help the worm swim. It is difficult to distinguish between the many species of paddleworms within the family Phyllodocidae. Definite identification requires very close examination. The gelatinous green egg capsules shown in the small photograph on page 125 were laid by paddleworms. [Up to 60 cm long]

Commensal ragworm - *Neanthes fucata*

Worm reaching for food among hermit crab mouthparts

Hermit crabs and the mollusc shells they inhabit often play host to a variety of animals. Some are obvious like sea anemones (pages 54 to 57) while others will only be seen with closer observation. Patient watching of a large common hermit crab (page 105), particularly when it is feeding, may result in a much rarer treat: the delightful sight of one of these ragworms emerging from its host's shell to share the meal. The larger photograph shows typical worm behaviour, where it is reaching round to put its head among the mouthparts of the crab. The smaller photographs show a worm reacting in apparently great excitement as its host arrives at the carcass of a dead spider crab – what a meal is in prospect! The worm's colour is generally pale brown but variable, with a distinctive white stripe that surrounds the main blood vessel running along its back. The worms are occasionally found free-living but youngsters will apparently wait in tubes on the seabed until they detect a hermit crab in whose shell they can take up residence. There, they build a mucus tube for themselves and then get protection and access to an excellent food supply. Any benefit the hermit crab gets from its lodger appears uncertain, although a cleaning function has been suggested. [Worm up to 20 cm long]

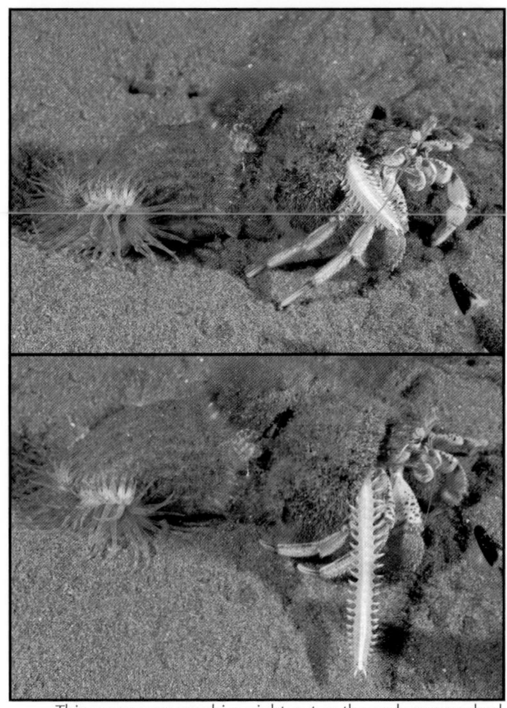

This worm was reaching right out as the crab approached food. Several animals, including a parasitic anemone (page 54), are living on the outside of the whelk shell occupied by the crab

Lugworm - *Arenicola marina*

Worm cast

Whether on the beach or underwater, the cast of the lugworm is a very familiar sight. The worm itself is hardly ever seen, as it has no need to leave its U-shaped burrow. Lugworms feed on the organic material in muddy sand and have to swallow enormous amounts of sediment in order to obtain adequate nutrition. The coiled casts represent processed material ejected by the worm while the small hollow in the sand nearby is where water, for respiration, and fresh sediment is drawn down. The long burrow is lined with mucus to prevent its collapse. [Worm up to 25 cm long; coiled casts up to 20 cm across]

Honeycomb worm - *Sabellaria alveolata*

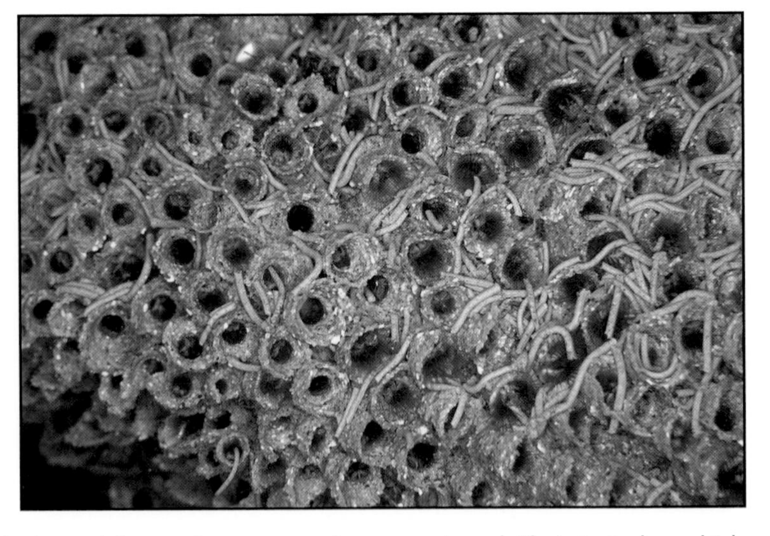

These worms, and all the other species up to page 138 in this chapter, lead sedentary lives in tubes that they make themselves (the lugworm's home is a burrow rather than a tube). The honeycomb worm is so called because the tubes of many individuals are usually grouped together to form a honeycomb-like reef structure. Their planktonic larvae can detect existing colonies and then settle on them. The honeycombs of tubes, produced by cementing sand grains together, are surprisingly fragile and should be treated with care. The reefs built by honeycomb worms are found draped over rocks amongst sand. Their tentacles, which can be seen most clearly in the top right of this photograph, trap floating food. [Worms up to 4 cm long, but reefs can look as though they are built by larger creatures]

Strawberry worm - *Eupolymnia nebulosa*

Strawberry worm found completely out of its tube

The long tentacles of a strawberry worm encounter a bootlace worm

Long sticky tentacles that spread out over the seabed to collect food particles are normally all that is seen of this worm. The tentacles shrink back quickly if touched, but are not fully retractable. Where many worms live close together, a real tangle of tentacles may cover the seabed. The body of the worm, and the slimy tube in which it lives, are usually hidden beneath stones and shell fragments on muddy sediment. Very occasionally, a worm may be found completely out of its tube (see main photograph) and the reason for its name is then obvious; the body is orange-pink with white spots. The smaller photograph shows a strawberry worm whose body was partly exposed. Its tentacles had encountered a bootlace worm (see page 127) and seemed to be investigating it; although there was no chance of such a large item being eaten! [Body up to 15 cm long, tentacles 20 cm long or more]

Sand-mason - *Lanice conchilega*

The exquisite architecture of a sand-mason's tube

Sand-mason - *Lanice conchilega*

Forest of sand-mason tubes

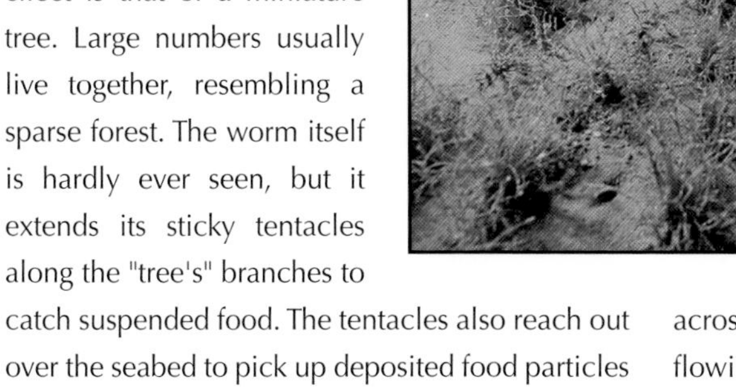

The sand-mason worm builds a gritty tube from grains of sand glued together with mucus (see opposite). The top of the tube stands proud of the seabed, and is crowned with a tuft of finger-like extensions, so the overall effect is that of a miniature tree. Large numbers usually live together, resembling a sparse forest. The worm itself is hardly ever seen, but it extends its sticky tentacles along the "tree's" branches to catch suspended food. The tentacles also reach out over the seabed to pick up deposited food particles and any sand grains required for tube maintenance. The branches, which are orientated across the current, also serve to slow down water flowing past, so more suspended material collects around them. [Body up to 30 cm long, tube to 45 cm but only top 5 cm shows]

Eyelash worm - *Myxicola infundibulum*

This worm's distinctive tentacles are webbed along almost all their length, so the crown resembles a small cone or funnel. It is just the flat, pointed and darkly coloured tips of the tentacles that are clearly separate. The eyelash worm lives in sandy and muddy areas and only the very end of its thick gelatinous tube, if any at all, protrudes from the sediment. The outer edges of the crown will be virtually flat on the seabed when fully expanded. This species is very shy and may rapidly withdraw into its tube when approached. The photograph shows a group of three individuals, but they are usually seen singly. [Tentacle crown up to 5 cm across]

Double spiral (or fan) worm - *Bispira volutacornis*

This species of fan worm is most abundant in the south, where its attractive feathery tentacles are commonly seen emerging from the nooks, crannies and overhangs along the sides of rocky gullies and reefs. The worm lives in a tube, constructed from mucus and mud, that looks like a roll of thick grey paper. Tubes may be up to 20 cm long, but most of their length is usually hidden within the rock crevice. When extended, the all-white or brown-and-white banded tentacles form a double spiral, hence the name. The tentacles have a number of functions, including the collection of suspended food and the extraction of oxygen from the seawater. They are also scattered with sense organs such as eyes. A shadow passing overhead, or sudden movement anywhere in the vicinity, will cause them to be rapidly withdrawn into the tube. The worms tend to live in small groups and while the retraction of individuals may often appear synchronised, those in some groups show marked variations in sensitivity and response time. A tube with tentacles withdrawn, and showing the typical pinched appearance, is at the top right of the photograph. [Tentacle crown up to 5 cm across]

Peacock worm - *Sabella pavonina*

The peacock worm can be a very striking animal, possessing a wonderful feathery fan of tentacles that emerges from a prominent but slim tube. These tubes are found attached to stones in sand and mud, and also on rocks or shipwrecks. The tentacles are usually colourful, often with red banding, and as with all tube worms they disappear quickly back into the tube if not approached with care. Water-borne particles captured by the worm's tentacles are sorted so that the largest are discarded, the smallest are eaten and intermediates are mixed with mucus and used for tube construction. The photograph shows a group of peacock worms living on the steep side of Loch Duich in the west of Scotland. Many peacock worms, such as those found living amongst eel-grass in sandy estuaries, are less obvious because their tubes are partially buried in the seabed. [Tubes up to 25 cm long, tentacle crown up to 15 cm across]

Organ pipe (or red fan) worm - *Serpula vermicularis*

Close-up of *Serpula* tubes showing tentacles and trumpet-shaped plugs

Organ pipe (or red fan) worm - *Serpula vermicularis*

The organ pipe worm is an impressive but very wary animal and its beautiful spread of red, pink or white tentacles tends to shoot back into its home tube at the slightest disturbance. With patience, however, the tentacles are eventually seen to re-emerge and can be examined along with the trumpet-shaped structure that forms a plug to the tube when the tentacles withdraw. The rigid chalky tube is anchored at its base but much of its length may stand proud of the seabed. Several tubes are often entwined around each other and, in sheltered water locations such as some Scottish sea lochs, large masses of worm tubes can be found (see smaller photograph). These living reefs provide shelter and attachment points for a huge number of other creatures such as brittle stars, crabs, scallops, sea squirts and clingfish to name a few. [Worms up to 7 cm long, masses of tubes can be nearly 1 m tall]

Spirorbis spirorbis

This worm lives permanently in a small coiled tube which is cemented on to the fronds of green or brown seaweed. Its favourite home is the saw wrack seaweed (*Fucus serratus*) that has distinctive serrated edges to its fronds. Many of the white chalky tubes are often found together. The worm has a crown of feeding tentacles that can be seen extended from the tubes of several of the worms in the photograph. One of the tentacles is specially modified to form a plug for the tube when the crown is withdrawn. A number of worm species build similar tubes; including other *Spirorbis* species that favour different habitats such as red seaweed or stones. [Tube coil is just a few mm across]

Coral worm - *Salmacina dysteri*

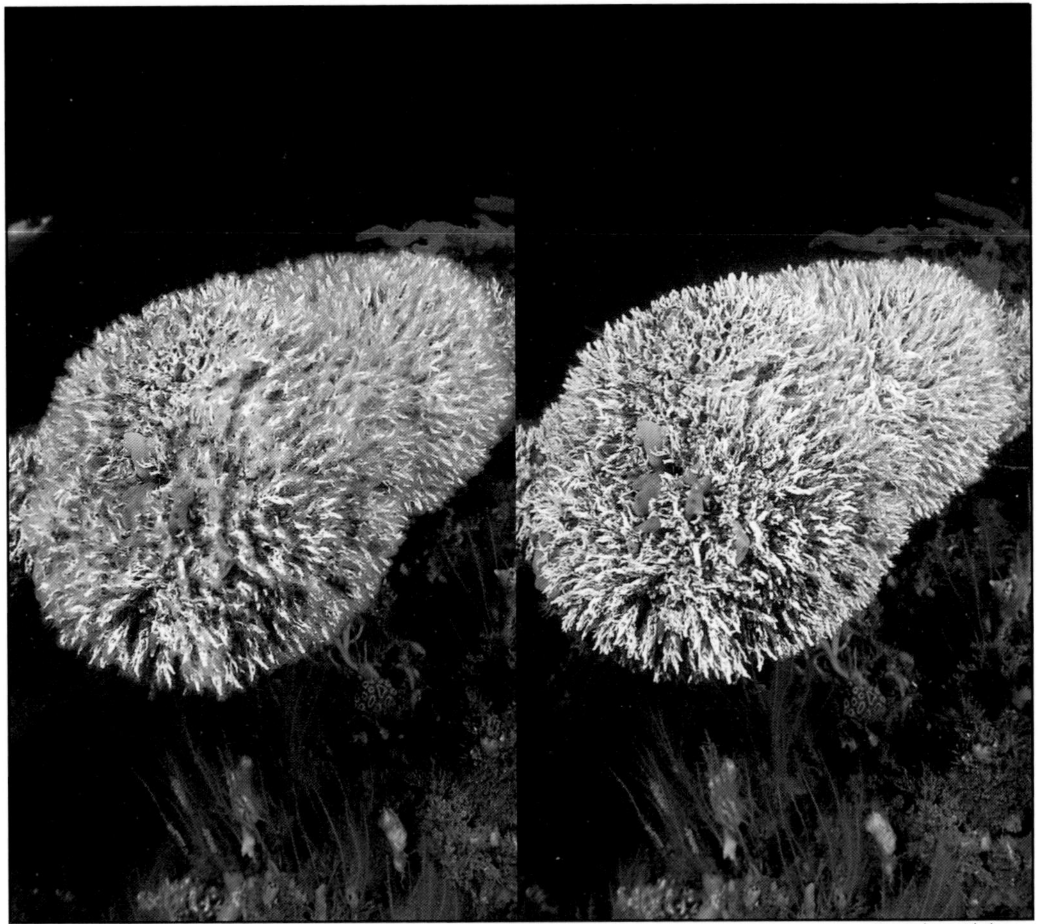

Tentacles out

Tentacles withdrawn

The coral worm is like *Serpula vermicularis* and *Spirorbis spirorbis* (pages 136-137) and different from the other tube worms described in this chapter, in that its tube is hard and completely rigid. Its name arises because the chalky tubes of many individuals are joined together to form a coral-like mass. Colonies can be found attached to any hard surface; the photographs show one that has encrusted an undersea cable. The fine tentacles of each worm give the colony a "furry" appearance when extended for feeding (left-hand photograph), but this changes dramatically when they are withdrawn (right-hand photograph). The coral worm differs from *Serpula*, *Spirorbis* and several other similar worms in that none of its tentacles are modified to form a special plug for the tube. The thin chalky tubes are manufactured almost entirely from calcium carbonate, the worms extracting calcium from seawater. Tube material is secreted by special glands, and is then moulded into shape by the collar near the worm's front end. Coral worms engage in some asexual reproduction (budding) and a new individual can be formed at the rear end of an existing one. The new worm will crawl part-way down the parent's tube and dissolve a small hole so that its own tentacles can emerge. If the parent tube is badly damaged by this activity it may become separated from the rest of the colony. [Individual worms tiny but colonies are often 15 cm across]

A marine leech - *Pontobdella muricata*

Leeches are not usually thought of as marine animals but, while most species live in freshwater or on land, some live in the sea. Leeches are of course primarily regarded as bloodsuckers and this is how most of them live, although some freshwater species are more conventional predators. The marine leeches are typical in being well adapted for their lifestyle as bloodsuckers on fish, with a sucker at the front surrounding the mouth and another at the rear so they can move across their host's skin by "looping" their body. *Pontobdella muricata* is a large and obvious leech, which specialises in parasitising skates and rays, and sometimes plaice. These ones were seen attached to a thornback ray in the shallows of Loch Kishorn. Other, usually much smaller, leech species in our waters target fish such as blennies, sea scorpions and angler fish. Leeches are classified as segmented worms (belonging to phylum Annelida) but are not in the bristle worm sub-group. All the other segmented worms on pages 128-138 are bristle worms. [*Pontobdella muricata* up to 20 cm long]

Leeches on a thornback ray (both photographs)

Chapter 6
MOLLUSCS
Chitons, sea snails, sea slugs, bivalves, cuttlefish, octopus & squid

The molluscs are perhaps the most surprising group of animals. The typical mollusc has a hard chalky external shell, formed in either one part (limpets, whelks) or two (mussels, scallops). Some molluscs, however, depart from this typical pattern. Those that have no shell at all (nudibranch sea slugs, octopus) or one that is hidden from view (sea hares, cuttlefish) are some of the most fascinating marine animals. Where present, the external shell is like a cover rather than the all-encompassing body armour of crustaceans. This means it can be gradually enlarged as the animal grows and does not have to be shed periodically, the same shell being kept for life. The phylum name, Mollusca, is derived from the Latin word for soft, in reference to the soft body typically enclosed in the hard shell.

Main mollusc features

The soft body of a typical mollusc is composed of three main parts: a muscular foot, a visceral mass which contains digestive and reproductive organs and, thirdly, the mantle tissue that secretes the shell. The very strong muscular foot forms the base of the animal and represents most of its contact with the outside world. The head, where sensory organs such as eyes and tentacles are located, is formed from the front end of the foot. The mollusc shell is made up of a protein matrix reinforced by numerous calcium carbonate (chalk) crystals to produce a strong composite material like fibreglass.

Unusual feeding machinery

An intricate feeding mechanism, the radula, is found in most molluscs (apart from the bivalves) but nowhere else in the animal kingdom. It is a ribbon of horny tissue, bearing teeth, which is drawn backwards and forwards across food like a file. It also acts like a conveyor belt in transporting the rasped off food to the digestive tract. The form of the radula varies between molluscs that browse vegetation and those that eat flesh. The hard working machinery is continually replaced, with new teeth produced at one end of the radula while worn teeth are broken off at the other end.

Reproduction

Molluscs use a great variety of reproductive strategies. Some simply shed eggs and sperm into the water while others use internal fertilisation and attach their eggs to the seabed. Still others brood eggs within their bodies and produce youngsters which resemble miniature adults complete with shell. Closely-related species (some of the periwinkles for example) may use very different strategies from one another.

Chitons

These small molluscs, also called coat-of-mail shells, can be easily overlooked. Their shell, instead of being a single structure as in the sea snails, is made up of articulated plates. Chitons are described on page 144.

Sea snails

These are the archetypal molluscs with a single hard shell, like limpets, whelks, top-shells and periwinkles. Many have a shell in the shape of a coil. They follow the basic mollusc body plan, as outlined in "Main mollusc features", quite closely.

Sea slugs

Few animals arouse such interest relative to their size as the colourful sea slugs. A small minority of sea slug species have an obvious shell, some have a much-reduced shell hidden by their soft tissues and the majority have no shell at all. Most of those with no shell belong to the nudibranch order. The term nudibranch actually means "naked gill" because the gills have no shell or mantle cavity to protect them. As virtually all sea slugs have no fully protective shell into which they can retreat, they have to rely on other forms of defence. Special skin glands produce toxins and irritants of various kinds to repel predators. An example of their effect is shown in the pair of photographs on this page. Some nudibranch species have an even more impressive system whereby they feed on creatures with stinging cells (sea anemones, hydroids) and utilise the second-hand cells for their own defence (see page 171). All sea slugs are hermaphrodites, having both male and female sex organs. When two come together to mate, there is usually double copulation with both individuals donating and receiving sperm. Sea slug egg masses are a very common sight attached to rocks, stones or hydroids. They are usually white and their appearance can vary from coiled flat ribbons to miniature strings of pearls. Only a fraction of the large number of sea slug species that can be seen are described in this chapter. For a full listing, one of the more specialist texts needs to be consulted (page 310).

A black goby nibbles an elegant sea slug (see page 166)

A few seconds later, the goby rejects it

Bivalves

These are molluscs with two halves to their shells such as mussels and scallops. Bivalves have no head and are the only molluscs to do without a radula. The name bivalve means "two shells" but the shell is actually a single structure. The narrow strip of shell joining the two halves has much less calcified reinforcement than the rest, so it can act as a flexible hinge. Though they appear static and rather unsophisticated, bivalve molluscs are highly specialised for their chosen way of life, namely filter feeding. The gills are enormous, far larger than would be needed just for respiration, because they are also used for collecting suspended food from the water. Numerous banks of cilia (tiny hairs) on the gills beat in unison to create a powerful water current through the body cavity while other cilia help to trap food particles and move them towards the gut. It is this filtering ability that can make bivalves dangerous purveyors of food poisoning when harvested from polluted areas. In addition to relatively obvious bivalves, such as mussels and scallops, there are numerous other types that burrow deep into soft sediment. At the surface of the mud or sand, the only sign of their presence is the openings of their water intake and outflow tubes. The tubes of large bivalves can be very obvious (see top photograph on this page) but they are withdrawn rapidly if any movement is sensed nearby. The other photograph of a sandy seabed on this page shows the much less obvious tubes (yellowish circles, a couple of mm across) belonging to small banded wedge shells. The shell in their midst is from an individual that was eaten by a necklace shell (see page 152); the hole drilled by the predator is clear to see.

The tubes of a large burrowing bivalve visible at the sand's surface

Banded wedge shell drilled and eaten by necklace shell, with the tube openings of living individuals nearby

Cuttlefish, octopus and squid

These fascinating animals make up a specialised group of molluscs, the cephalopods. The muscular foot of typical molluscs has become the group of tentacles attached to the head (cephalopod means "head-foot") and the mantle tissue has formed a jet propulsion organ, but it is still difficult to visualise how the basic mollusc body plan has been adapted to produce such sophisticated animals. Rather than having separate nerve centres scattered round the body like other molluscs, these animals have them fused and enlarged to form a sophisticated brain which is enormous by the standards of invertebrates. [Invertebrates are animals without proper backbones - all the animals in this book, except for the fish, are invertebrates]. The cephalopod brain is responsible for sophisticated behaviour and these animals display the ability to learn and remember for several weeks. Cuttlefish and octopus are described in detail on pages 180 to 189. Squid are rapid open-water swimmers and are rarely seen underwater. The common squid shown below was dying, having possibly just reproduced. Their distinctive egg clusters, which look like long white sausages, are generally seen more often than the animals themselves. The bottom photograph shows squid eggs attached to a pink sea fan.

Dying common squid

Squid eggs on a pink sea fan

Chitons

The very characteristic shell of chitons (coat-of-mail shells) is made up of eight arched plates which interlock along their edges. A thick rim of leathery mantle tissue, known as a girdle, is visible around the outside of the shell. Chitons are considered to be relatively primitive molluscs, with features similar to those of the ancestral molluscs from which all the other groups are descended. Their flattened shape and articulated shell makes them very well adapted for clinging tightly on to (often underneath) rocks or stones. Found mainly on the shore or in shallow water, they crawl very slowly over the rocks, grazing on algae and small encrusting animals such as bryozoans. Chitons have only a simple nervous system and no tentacles or eyes, although some are believed to carry simple light detectors in their shell plates. About a dozen species of chiton live in British waters, and two examples are shown here. *Lepidochitona cinerea* is common all around our coasts while *Tonicella marmorea* is a northern species, rarely found further south than North Wales on the west coast and Yorkshire on the east. Identification of different chiton species requires close examination of the shape of their shell plates and the features on their girdles.

Lepidochitona cinerea, 2 cm long

Tonicella marmorea, 3 cm long

Blue-rayed limpet - *Helcion pellucidum*

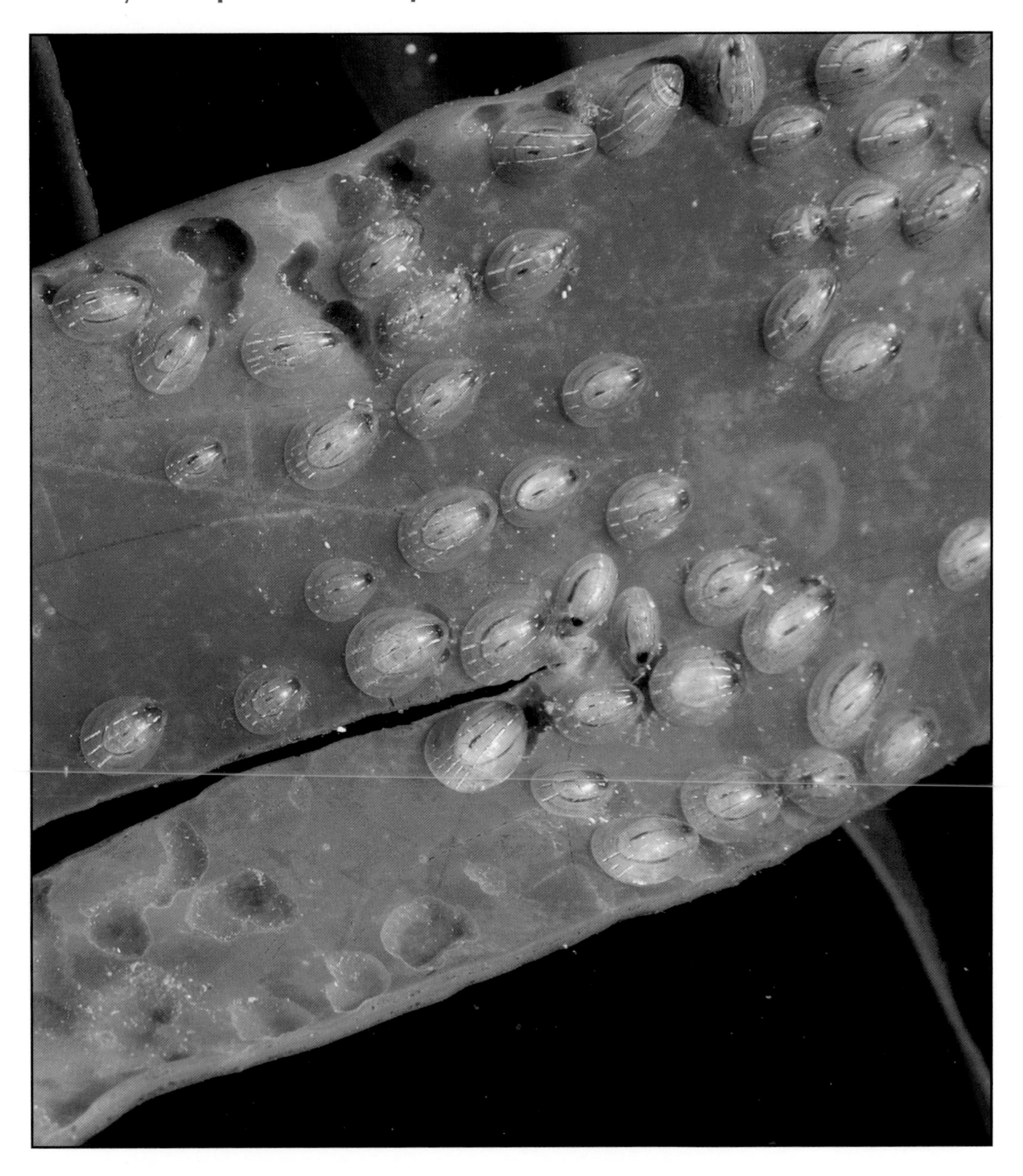

This beautiful little animal is only found on kelp plants, so it is often overlooked by divers and snorkellers to whom the swaying forests of kelp appear relatively unappealing. The delicate, slightly translucent shell is usually kelp coloured and would be unremarkable if it was not marked by broken lines of a wonderful kingfisher blue. The limpets are often found in small groups and each one excavates its own little pit into the kelp frond or stalk as it feeds. Some of these limpets appear to move down the kelp as autumn approaches, so they avoid being cast adrift when the frond is lost or damaged in winter storms. A different variety of the same species has a rougher shell with less prominent blue markings, and lives down in the holdfast of the kelp. [Shell up to 2 cm across]

Limpets - (various *Patella* species)

Limpets belonging to the genus *Patella* are amongst the most familiar molluscs and are abundant on virtually all rocky shores. They have a robust conical shell that is usually seen jammed down tightly onto the rock surface where they live. If one is found while it is actively crawling along (see top photograph) the muscular foot and a pair of large head tentacles may be visible, along with numerous tiny tentacles around the very edge of the shell. Like so many animals that seem mundane at first glance, limpets have an interesting story to tell. They create a "home base" on the surface of a rock by grinding their shell (or the rock if it is soft) to make a perfect fit. This helps to prevent the animal drying up when the tide is out. When the tide is in, limpets leave their "home base" (visible as a scar on the rock, see bottom photograph) and go off to graze algae on the rock nearby. They have to return home before the tide goes out

Limpet crawling

Submerged rock at high tide showing circular limpet scars

Limpets - (various *Patella* species)

Limpets grazing in the shallows at high tide; two limpets in the centre of the photograph appeared to be in dispute

again, and they navigate by following the trail of mucus which they left on the way out. Some sort of memory has also been suggested. Rocky shallows at high tide with limpets grazing (see main photograph this page) may seem a very peaceful scene but, if you watch closely, disputes between two limpets can be observed. This aggression presumably stems from competition for grazing territory. It involves one limpet ramming its shell into another's and often progresses to the aggressor shoving the edge of its shell underneath the lip of the other and then lifting, as though trying to prise it from the rock. An example of this, surprisingly seen at low tide (with just a thin film of water remaining on the rock) is shown in the smaller photograph on this page. The dispute continued for about 30 minutes but there was no obvious outcome.

Aggression between two limpets, surprisingly seen at low tide

Large limpets also have an intriguing method of defending themselves against predatory starfish. They will apparently lift up their shell and slam the edge down on the starfish's arm or tube feet. The three different species of *Patella* (*P. vulgata* the common limpet, *P. ulyssiponensis* the china limpet and *P. depressa* the black-footed limpet) are difficult to distinguish without removing them from their rock and this can maim or kill them. [Shell up to 6 cm long]

Painted top-shell - *Calliostoma zizyphinum*

The shell of this mollusc is in the shape of a sharply-pointed, straight-sided cone. Shell colouration is usually an attractive yellowish pink with streaks and blotches of crimson or brown. Completely white individuals may also be found. The animal apparently keeps its shell clean by regularly rubbing it with its extendable foot. Like all top-shells, *Calliostoma* is a grazer and feeds on algae and other tiny organisms that live on the surfaces over which it crawls. In the summer, it can sometimes be found laying its long gelatinous egg ribbon which it attaches to stones or rock. This photograph shows a top-shell crawling up a kelp stem. The animal's foot, with its associated tentacles extending up onto the shell, can be seen. [Shell up to 3 cm across]

Purple (or flat) top-shell - *Gibbula umbilicalis*

Grey top-shell - *Gibbula cineraria*

Purple (or flat) top-shell   Tentacles visible in both photographs   Grey top-shell

These small top-shells can often be seen in good numbers grazing on seaweed and rocks on the shore and in the shallows. The purple (or flat) top-shell, distinguished by a flatter shell and obvious, well-spaced pinkish purple stripes, is found from the upper shore down to low tide mark. The grey top-shell has a taller shell which is greyish overall with numerous narrow wavy reddish lines that may not always be clear. It is found from the lower shore down into shallow water, where it will most easily be spotted grazing on kelp fronds. Other top-shells can be seen too, and it is well worth consulting a specialist shell or shore guide (see page 310) for more comprehensive information. [Shell of both purple and grey top-shells up to about 2 cm across]

Tower (or auger) shell - *Turritella communis*

These very distinctive long, pointed and screw-like shells are often found empty and strewn across a sandy seabed; or in use as the jauntily-carried homes of small hermit crabs. When the original mollusc owner is in residence, the shell lies buried just under the surface of muddy sand where its shape helps it to stay screwed into position. The animal is well adapted to this sedentary lifestyle and simply pumps in seawater from which it extracts oxygen and filters out particles of food. A cluster of the tower shell's egg capsules can be seen in the photograph as sandy-coloured beads attached to their stalk. To the left of the picture are a sand-mason worm (page 132) and a burrowing anemone (page 59). [Shell up to 5 cm long]

Edible periwinkle - *Littorina littorea*

Edible periwinkles are widespread and found in huge numbers on the shore and in shallow water. Usually thought of as rocky coast animals, they also live in muddy areas such as estuaries. The shell is shaped like that of a land snail and is coloured black, grey or brown. Several similar types of periwinkle, including other species of the *Littorina* genus (see example below) are also very common. Their identification can be complicated and a specialist shell or shore guide (see page 310) is again recommended. All periwinkles are herbivorous grazers but the life histories of the different species vary greatly (see more details below). [Edible periwinkle shell up to 3 cm across, most other types are smaller]

Flat periwinkle - *Littorina obtusata* and *Littorina mariae*

Flat periwinkles are common on the shore amongst seaweed. They are small and have fairly smooth shells with, as the name suggests, very much flattened spires. Shell colour is highly variable with yellow, olive green and brown forms all being common; the colours found differ between habitats. This is thought to be a result of natural selection, with the particular environment determining which coloured shells are most obvious to predators. The two species of flat periwinkle (*Littorina obtusata* and *Littorina mariae*) are difficult to distinguish and this is usually done by close examination of the shell opening shape although, even with this, the separation can be unclear. Before 1966, they were regarded as a single species. Flat periwinkles graze on seaweed, on which they also lay their eggs which hatch into miniature crawling versions of the adults. This contrasts with the edible periwinkle (above) which releases its fertilised eggs into the sea where they hatch to produce planktonic larvae. Other periwinkle species even brood their eggs within their bodies and give birth to fully-formed young, showing how these closely related species can have surprisingly different breeding habits. [Flat periwinkle shells up to 1.5 cm across]

Slipper limpet - *Crepidula fornicata*

Shell opening with shelf, producing the "slipper" shape

Now widespread around the south of Britain, this species was accidentally introduced from America back in the 19th century. It arrived in a shipment of oysters which were then re-laid for fattening on British oyster beds. The shell is a flattish oval shape and has a large shelf across its opening, visible when turned over (see top photograph), which gives it the appearance of a slipper. Its most remarkable feature is the way in which it forms stacks of up to twelve individuals (the bottom photograph shows a stack of six). Larger stacks tend to be curved like an arch, hence the specific name of *fornicata*, Latin for arch. The large limpets at the bottom of a chain are female, the small ones at the top are male and the ones in between are, as one might suspect, in between. The animals change from male to female as they grow older. A young slipper limpet settling on its own develops into a female relatively rapidly, while one settling at the end of a chain will spend some time as a functional male. It appears that females secrete a sort of hormone into the water which maintains the masculinity of nearby males. The adult animals are immobile but males use their long penis to fertilise a female within the chain. Where there are dense masses of slipper limpets, near Weymouth for example, they seem to take over the sea floor. They are filter feeders and can be a serious pest to oyster beds through crowding and competition for planktonic food. [Shell up to 5 cm long]

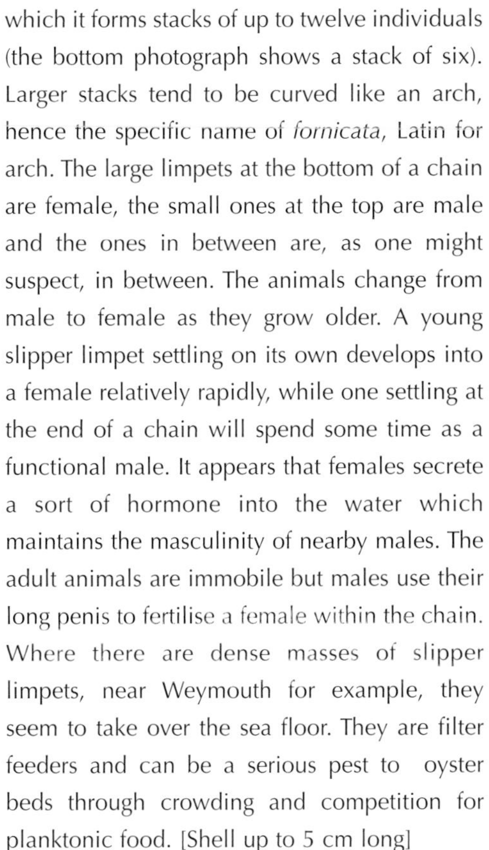

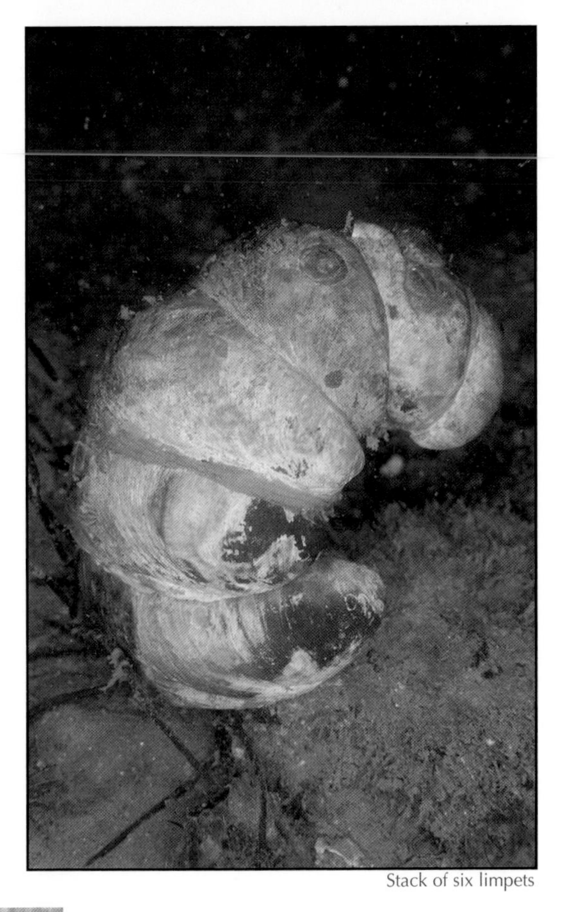

Stack of six limpets

Large necklace shell (or moon snail) - *Euspira catena*

The foot of this attractive mollusc spreads up onto its shell when it is active, and also out across the sand like a skirt as it crawls along. Its rounded shell is glossy and an overall pale orange or fawn, with a row of brownish streaks running around the upper part of the spiral. The necklace shell preys on small clam-like bivalve molluscs that are buried in the sand. Having found the bivalve's siphons at the surface of the sand, it burrows down to attack. The bivalve shell is first softened chemically and then bored through with the necklace shell's drill-like radula. This leaves a neat round hole near the hinge that can often be seen on empty bivalve shells washed up on the beach or lying on the surface of the sand underwater (see page 142). The spreading of the necklace shell's foot up around its shell makes it into a more streamlined form for burrowing. It is also useful in defence, as the tube feet of a predatory starfish cannot get a grip on the slimy tissue as easily as they could on a bare shell. Necklace shells lay a distinctive egg ribbon that looks just like a collar (see inset photograph). Most of the eggs in the ribbon end up feeding the small proportion of embryos that develop. This species is found on sandy seabeds all around Britain, at virtually any depth. [Shell up to 3 cm tall]

European cowrie - *Trivia monacha*

The cowrie's shell is shiny and appears highly polished but, in an active animal, it is almost completely obscured by folds of soft mantle tissue which extend up from the entrance in the base of the shell. These folds have a rather striking pattern, a little like that of fake leopard-skin. The pale shell is traversed by delicate ribs and there are three dark blotches along its top (the blotches are absent on the shell of the Arctic cowrie, *Trivia arctica*, a close relative). The European cowrie is more common in the south and west of Britain, with the Arctic cowrie more so in the north. Both species of cowrie eat colonial sea squirts, also laying their eggs in holes bitten out of the colonies. The photograph shows a European cowrie crawling across sand, but they are more commonly found on rocks and stones. [European cowrie, shell up to 1.5 cm long. Arctic cowrie, shell up to 1 cm]

Dog-whelk - *Nucella lapillus*

Dog-whelks are abundant on rocky beaches. Though small, the shell is quite heavily built to withstand the rigours of living on wave-battered shores. It usually bears obvious spiral ridges but these may be worn virtually smooth in some animals. Shells are generally white or pale grey but differently coloured and attractive banded forms (top photograph, page 154) can also be seen. The dog-whelk is an **active carnivore**, forcing open the top plates of barnacles or boring through the shells of mussels to feast

Preying on barnacles

Dog-whelk - *Nucella lapillus*

on their soft insides. The boring process, which uses a combination of chemical softening and mechanical drilling or rasping, can take several hours or even days. A mussel is not, however, always the helpless victim. While a dog-whelk is preoccupied with its drilling, the mussel and its neighbours may attach so many of their byssus threads (like miniature guy-ropes, see page 174) to the dog-whelk that it is turned over or at least **immobilised**. It will then eventually starve to death (middle photograph, this page). As in most sea snails, the sexes are separate. Dog-whelks lay their eggs in small pale yellow bottle-shaped **egg capsules** (bottom photograph, this page) which can often be seen in crevices amongst rocks on the shore. The vast majority of up to a thousand eggs in each capsule are used as food by the fortunate few that develop. These youngsters hatch as miniature crawling dog-whelks. The dog-whelk's claim to fame in recent years has been as a **sensitive indicator of pollution** caused by tributyl tin (TBT) anti-fouling paint, which is used to prevent growth of encrusting creatures on boat hulls. Minute concentrations of TBT in seawater cause female dog-whelks to develop male sexual organs and become sterile, causing severe damage to some populations. Legislation on the use of TBT has allowed some recovery. [Shell up to 4 cm long]

Banded shell

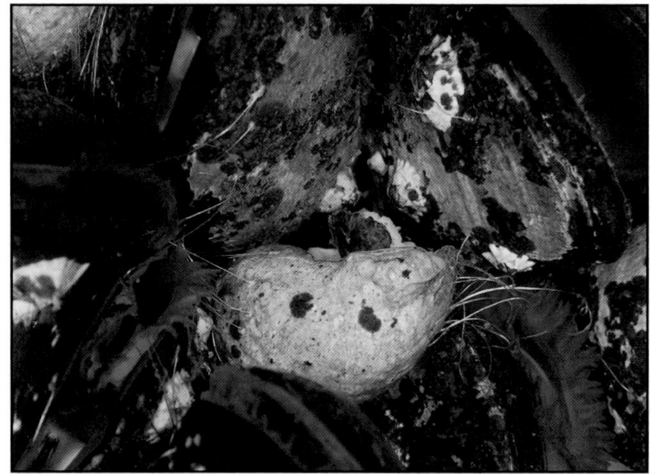

Tied down by mussel byssus threads and dying

Dog-whelk and egg capsules

Netted dog-whelk - *Hinia reticulata*

This animal is a very common sight on sand, particularly where there are rocks nearby. It has a conical shell, distinctively marked with a neat rectangular pattern (hence the name) formed by the interaction of flat ribs and spiral ridges. A long siphon is usually held aloft and draws water down to the animal for respiration, even when the shell is completely buried in the sand. The incoming water is also tested for evidence of carrion and these animals can detect food at a considerable distance. They are often found gathered in large numbers on a fish or crab carcass (see middle photograph) and a search of the sand nearby usually reveals several latecomers hurrying to the feast. The female lays very distinctive egg capsules in rows on eel-grass (see bottom photograph), seaweed or stones. They are shaped like tiny flattened vases and are transparent so the eggs can be seen inside. [Shell up to 3 cm long, egg capsules 0.5 cm tall]

Netted dog-whelks feeding on a crab carcass

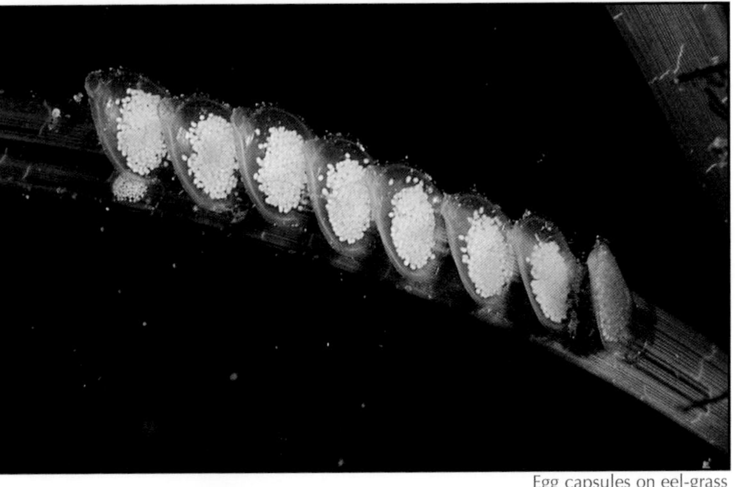

Egg capsules on eel-grass

Common whelk - *Buccinum undatum*

Whelk on the move, with siphon and tentacles extended, and shell "door" visible on the back of the foot

Left inset: common whelk laying eggs. Main photograph: bunch of egg capsules washed up. Right inset: shell "door" in use

A "snail on steroids" might be an apt description for this hefty mollusc that is usually found on soft seabeds. The spiral shell is heavily built and covered with lines and ribs; while the body is whitish with often obvious black flecks. The back of the whelk's foot carries a flat oval of horny material that looks surplus to requirements when the animal is crawling along. When the whelk is threatened and retreats into its shell, the function of this item can be seen as a neat lid or door that bars the shell's entrance (see right inset, above). Another method of self defence is employed when

Common whelk - *Buccinum undatum*

a whelk moves its body with a violent rocking motion to dislodge an attacking starfish. After mating, the female whelk will lay a mass of up to 2000 egg capsules attached to the seabed (see left inset, opposite). Each capsule contains approximately 1000 eggs. The vast majority of eggs and embryos fail to reach maturity and are used as food by the lucky few who survive. The sandy-coloured bunches of empty egg capsules are often found washed up on the shore (main photograph, opposite bottom) and are called "sea wash balls" because they were once used like a sponge or flannel. Adult whelks eat sand-dwelling worms as they crawl across the seabed and also prey on bivalve molluscs, such as cockles, by forcing the shells apart with the edge of their own shell. They are also voracious consumers of dead and dying animals which, as with the netted dog-whelk, can be detected from a good distance. Small hermit crabs are sometimes seen "riding" on the back of whelks (photograph above); could this be an easy way for them to find a carrion meal? [Shell up to 11 cm long]

Red whelk - *Neptunea antiqua*

This is a large animal similar to the common whelk. Differences are that the shell is smoother with less pronounced ridges and the very front part of the shell, through which the groove carrying the siphon runs, protrudes to a greater extent. Shell colour tends to be reddish as the name suggests but, confusingly, can also be very pale. This species is recorded as being most common off the west coast of Scotland and I see it frequently near Oban, where these photographs were taken. A small dragonet (page 303) can be seen under the rear of the red whelk's shell in the top photograph. The bottom photograph shows eggs being laid in a cluster similar to the common whelk's. [Shell up to 11 cm long]

Red whelk laying eggs

Acteon tornatilis

Despite having an obvious and robust-looking shell into which it can withdraw, this mollusc is classified as a sea slug. The glossy light pink and brown shell has a distinctive shape, a little like a rugby ball, and usually sports broad pale or white bands. Found in sheltered sandy bays, *Acteon* preys on burrowing worms, particularly sand-masons (pages 132). When it finds one of their tubes, it burrows down and grabs the sand-mason worm by its tail before proceeding to gradually eat it. The whole process may take several hours. The animal in the top photograph appears to be in the process of laying eggs. The bottom photograph shows a completed *Acteon* egg mass. [Shell and egg mass both up to about 3 cm long]

Egg mass

Scaphander lignarius

The thick glossy shell of this sea slug is neither large enough to accommodate the whole animal when it retracts nor small enough to be totally hidden by the white soft tissues. The animal burrows through muddy sand at a depth of up to 5 cm looking for prey such as worms, bivalve molluscs, small sea urchins and crustaceans. This individual, found on the surface of the sand, ploughed itself rapidly underneath it and out of sight just after the photograph was taken. The slug can release a viscous yellow liquid if it is disturbed. [Up to 6 cm long]

White mud slug - *Philine aperta*

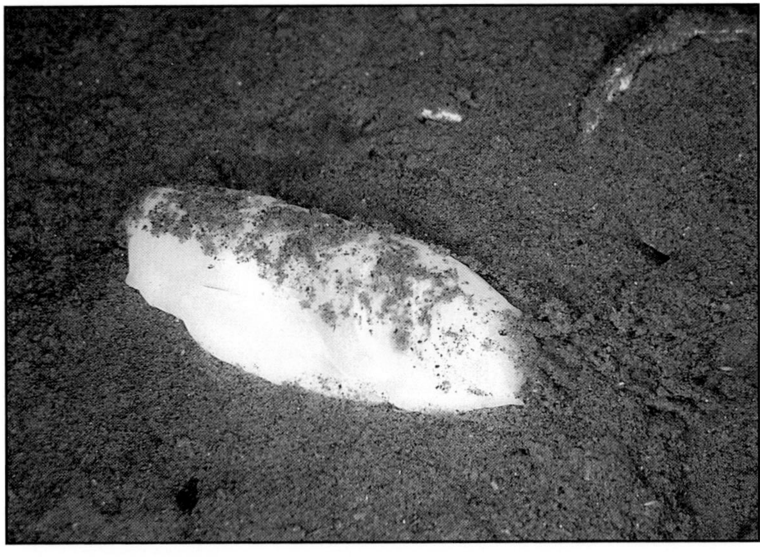

At first glance, this slightly nondescript looking sea slug can be mistaken for a piece of white debris lying on the seabed; but a trail left behind it in the form of a groove in the mud is often a sign of vitality. A closer look will reveal a soft and glossy white body that resembles a squashed egg. Like all the sea slugs on pages 158 to 162, this species is not a nudibranch. Its soft tissues hide an internal shell. *Philine aperta* can be abundant in muddy or sandy areas, where it burrows in search of worm and small mollusc prey. Although its skin produces sulphuric acid as a defensive reaction to ward off predators, it is thought to be eaten by haddock and flatfish. [Up to 7 cm long]

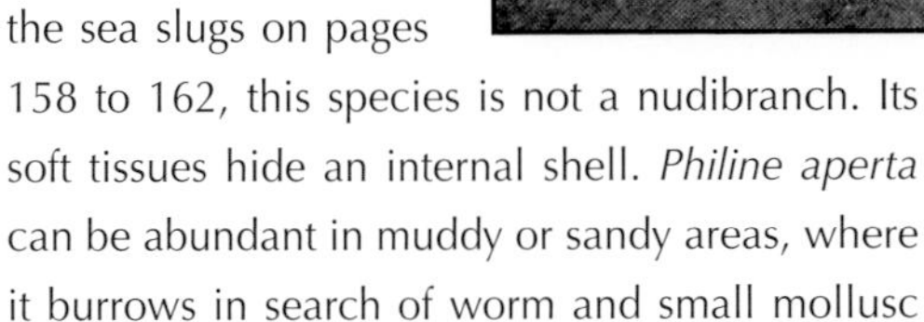

Akera bullata

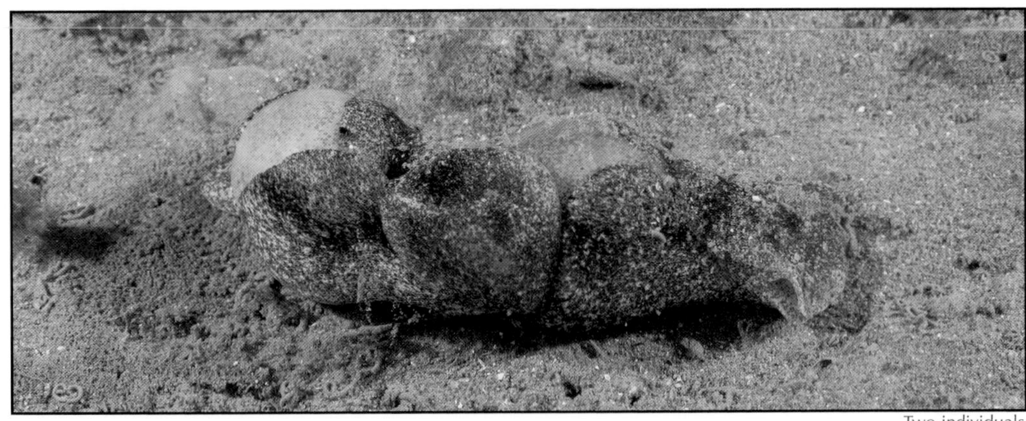

Two individuals

The colouration of this strange-looking sea slug makes it quite distinctive, although its body features can be hard to make out; as in this photograph of a pair that appeared to be about to mate. Dense dark mottling on the body largely covers the background colour of pale grey or brown. The glossy amber-coloured shell (for this is another sea slug that has one) can be seen between the large flaps of tissue at the rear of the body, as in both the individuals here. These flaps propel the slug when it swims, rather like wings. *Akera bullata* lives in sandy and muddy habitats, often among eel-grass which is where I have always seen them. It is a herbivore and is thought to feed on the eel-grass itself. [Up to 7 cm long]

Sea hare - *Aplysia punctata*

A small mating chain of only three sea hares

Sea hare - *Aplysia punctata*

The sea hare gets its name from the broad upper pair of head tentacles that look like hare's ears. It has large flap-like lobes on its back and there is a much reduced shell hidden by the soft tissues. Sea hares may be brown, olive-green or dark red, probably depending on the type of seaweed they eat. Individuals eating the appropriately named food of sea lettuce are thought to be green, while those eating red seaweeds (usually younger sea hares) can be almost maroon. A noxious mixture of foul-tasting whitish slime and vivid purple fluid, each produced by separate glands, is secreted for defence. The sea hare is a hermaphrodite like all sea slugs and any individual can act as either a male or female to another. Several sea hares will sometimes form a mating chain where each acts as a male to the one below it and as a female to the one above.

Large mating chain and pink egg string

The small photograph above shows a large chain, in which the individuals are difficult to discern, and also a pink strand of their eggs. The bottom photograph opposite shows a chain of just three. It is much easier to see what is happening, but only the middle sea hare is gaining the full experience of acting as male and female simultaneously. [Up to 20 cm long but usually much smaller]

Green sea slug - *Elysia viridis*

Like the sea hare, this attractive small sea slug is a vegetarian whose body colour is thought to depend on the type of seaweed that it eats. Usually green, it can vary to red, so the characteristic small spots of blue, green or red iridescence are a very useful identification feature. As in the sea hare, there are lobes or "wings" that may be either spread out or, as in this photograph, folded up over its back. It is different from the sea hare and the other non-nudibranch sea slugs described on pages 158 to 162, because it has no shell at all. [Up to 4 cm long]

Highland dancer - *Pleurobranchus membranaceus*

Laying egg ribbon

Swimming

This large sea slug has a pale brown body, usually with patches of darker brown, which is covered in soft tubercles that give it a "warty" appearance. Its mantle tissue forms a skirt around the body. The slug shown in the main photograph was in the midst of producing its coiled egg ribbon which may contain more than one million eggs. This species feeds on sea squirts by drilling through their tough outer tunic and sucking out the soft insides. Like the sea hare and white mud slug, it has an internal shell hidden by the soft tissues. With no protection offered by the shell, potential predators are deterred by sulphuric acid which is produced by the skin and released if it is broken. The highland dancer can swim well, always in an upside-down position, using undulating movements of its foot (smaller photograph). Very occasionally, huge populations have been witnessed undertaking swimming migrations. The purpose of these is a mystery but aggregation for mating seems to be the most likely reason. [Up to 12 cm long]

Dead men's finger sea slug - *Tritonia hombergi*

Egg mass amongst tube worms and sea anemones

This sea slug, and all those that follow in this chapter, are nudibranchs ("naked gills", see page 141) and lack a shell entirely. Despite being the largest nudibranch in British waters, *Tritonia hombergi* can be surprisingly difficult to spot. The numerous branched protrusions (gills) around its white or pinkish brown body help it to blend in with its surroundings, particularly as it is usually found amongst colonies of dead men's fingers soft coral (pages 66-67) on which it feeds. It is very selective in its choice of prey, like most nudibranchs, and is thought to eat nothing else. When away from soft corals on a contrasting seabed, as shown here, the striking appearance can be fully appreciated. The rounded bumps on its back produce an irritant compound that has been found to cause skin blisters on some people handling it. The eggs are laid as a twisted string which is attached to the seabed with a transparent membrane (see smaller photograph). This species, like all British sea slugs for which there is information, lives no longer than a year. [Up to 20 cm long]

Lomanotus genei

Mating pair

Lomanotus genei is a wonderfully flamboyant nudibranch. Its colour can vary from white through pink and orange to red, but it is characterised by a distinctive frill of protuberances with yellow or orange tips along the body. The main photograph shows a mating pair. As is typical with nudibranchs, the pair manage to both act as male and female during mating, thus fertilising each other simultaneously. The smaller photograph shows a single animal for ease of recognition. *Lomanotus* feeds on hydroids and winds its egg string around them at spawning time. It can swim clumsily if disturbed. [Up to 9 cm long]

Crystal sea slug - *Janolus cristatus*

This species, while not as colourful as some of its relatives, is still particularly striking. The pale translucent body is covered by numerous finger-like projections that appear rather bloated. A thin dark thread of digestive gland runs down the centre of each projection, sometimes forming tributaries near its end. There are splashes of an iridescent bluish-white pigment on the tips of the projections, and also in patches on the rest of the body. It is a fairly widespread animal, but is thought to be restricted to calm water because of its fragility. Its eggs look like a wavy string of tiny white beads reminiscent of a miniature pearl necklace; each bead contains around 250 individual eggs. [Up to 8 cm long]

Goniodoris nodosa

This common small translucent white sea slug has an obvious crown of feathery gills on its back towards the rear. Other notable features include the small bumps and specks of pigment scattered over its body and the pale ridge, which runs down its back and tail". There is also a characteristic small transparent patch behind the gills that looks like a hole. Youngsters feed on bryozoans especially sea chervil (page 192) and adults on sea squirts such as the gooseberry squirt (page 236). This photograph shows an individual crawling past eel-grass stems on a sandy seabed. [Up to 3 cm long]

Elegant sea slug - *Okenia elegans*

The body of this beautiful nudibranch sea slug is basically white, but may have such dense red freckles that it appears pink. There is a brilliant yellow border around the base of the body and the gills and various body protuberances are also splashed with yellow. The pair of tentacles, known as rhinophores, on the upper side of the head are highly convoluted and patterned in red and white with pale tips. The elegant sea slug feeds on sea squirts by burrowing into them, so it is often only the tips of the feathery gills that can be seen. It is usually classified as a rare species but can be quite common in locations such as Plymouth Sound and Torbay. Most sightings come from south west Britain. [Up to 8 cm long]

Laying eggs

Onchidoris bilamellata

Aggregation of sea slugs with their egg ribbons (both photographs)

This species may appear drab in comparison to its more flamboyant nudibranch sea slug colleagues, but the large aggregations it can form are impressive in their own right. The slug's body is a basic off-white but there is usually a dense brown blotchy pattern on its back. Numerous small projections all over the body tend to show through as off-white bumps. The head bears a pair of obvious ridged head tentacles (rhinophores) and there is a crown of feathery gills at the rear of the body. These slugs feed on barnacles and can occur in dense gatherings where their prey is plentiful, such as on a seabed of barnacle-encrusted stones and boulders. The top "lid" plates of the barnacles are chewed away so that the soft inner body is sucked out, leaving an empty base. Because the empty bases are paler than intact barnacles, an area recently ravaged by these slugs has quite a distinctive appearance. The aggregations shown in the photographs were found near Oban in August; many of the individuals were mating and ribbons of eggs were scattered over the seabed. Acidic defensive secretions are produced by this species if it is disturbed. A closely related but smaller species, *Onchidoris muricata*, feeds on bryozoans and is shown on page 191. [*O. bilamellata* up to 4 cm long, *O. muricata* up to 1.5 cm long]

Fried egg sea slug - *Diaphorodoris luteocincta*

The colouration of this small nudibranch sea slug species is both conspicuous and distinctive. Against an overall background of white, there is a large red blotch down the centre of its back and a yellow rim around the edge of its body. It has a similar form to *Onchidoris bilamellata* (page 167) and, like that species, the bumps on its back appear white through the darker blotching. The fried egg sea slug feeds on particular species of bryozoans and is usually found on the silt-covered rocks where these thrive. The pair in the photograph had probably just mated or were about to mate. [Up to 1 cm long]

White hedgehog sea slug - *Acanthodoris pilosa*

Its "fluffy" appearance is a characteristic of this common species. This impression arises from the long and soft protrusions on its body surface, which contrast with the more rounded bumps of several other species of nudibranch. The head tentacles (rhinophores) are club-shaped and there is a large crown of gills at its rear. Colour can be white or any shade of brown through to black, or even purple. The usual prey of this slug includes the bryozoan (*Alcyonidium diaphanum*) known as sea chervil. The photograph (small, left) on page 192 shows the sea slugs and their eggs on a colony of the bryozoan. [Up to 5 cm long]

Yellow edged polycera - *Polycera faeroensis*

The translucent white body of this nudibranch (common in the south and west) tends to appear slightly swollen. Various parts of it are splashed with bright yellow, such as the distinctive tapering extensions around the front of the head (of which there are eight or more). There is also yellow on the tips of the pair of tentacles on top of the head, on the gills towards the rear, and on the protuberances next to them. A closely related and more widespread species, *Polycera quadrilineata* ("lined polycera"), is similar in appearance but has fewer tapering extensions at the very front of the head (usually four) and splashes of yellow or orange forming several lines down the body itself. [Up to 5 cm long]

Orange clubbed sea slug - *Limacia clavigera*

A small white nudibranch with yellow or orange marking on the bumps and finger-like protuberances round its head and down the back of its body. The more rounded shape and form of these protuberances distinguish this species from *Polycera faeroensis* (above). It feeds on encrusting bryozoans such as the sea-mat (page 191). The pair in the photograph were resting on a mesh-like sea-mat colony and were in the process of mating. It can be seen that the sexual organs on the right-hand side of the slugs' bodies are joined so that there is mutual fertilisation. [Up to 2 cm long]

Pair mating

Sea lemon - *Archidoris pseudoargus*

The sea lemon is a large nudibranch, very common all around Britain. The upper side of its body is covered in small wart-like bumps and usually bears blotchy markings which can be any combination of yellow, pink, white, brown or green. A ring of feathery gills sticks up from near the animal's rear but is retracted quickly on the sensing of any disturbance. Sea lemons feed exclusively on encrusting sponges, chiefly the breadcrumb sponge (page 31), and their seemingly garish colouration can actually make them extremely difficult to spot when they are crawling across a mass of sponge. The smaller photograph shows a sea lemon's egg ribbon, which is laid in a coil with its bottom edge attached to the substrate, so a characteristic rosette is formed. Another ribbon is being laid in the background. The sea lemon's entire life cycle is completed within a year. Adults mate and spawn in the spring before dying, and the new juveniles appear in the late summer and grow through the autumn and winter. [Up to 12 cm long]

Slug laying egg ribbon and completed rosette

Coryphella browni

The body of this nudibranch is a translucent white but the numerous long and pointed projections, or cerata, on its back contain brightly coloured (usually crimson) tributaries of the digestive gland. There is also a broad and very obvious white band near the tip of each projection. *Coryphella browni* feeds on hydroids such as the oaten pipe (page 74) and can often be found munching away at their polyps while perched on the straw-like stems (see photograph). Not only is it undeterred by their stinging abilities, it is one of the nudibranchs that ingests the stinging cells of such animals and puts them to its own use. The slug passes the intact cells through its digestive system and out to the cerata tips, where they are used for defence against predators. A hungry fish that attacks the slug is almost bound to nip the cerata, causing stinging cells to discharge and the attack to be abandoned. Egg masses of this species, which look like wavy white threads, are often laid amongst the hydroid stems. Several other nudibranch species, some of which belong to the same *Coryphella* genus, have a very similar appearance. [Up to 5 cm long]

Coryphella lineata

This beautiful sea slug is probably the most easily identified of the various *Coryphella* species because of its distinctive markings. There is typically a thin white pin-stripe that runs down the middle of the back and forks into two on the front (oral) pair of head tentacles; and a similar line down each flank that join the mid-line back at the tail. There is also a line down each of the projections (cerata) on its back, but it is the whole-body pin-stripes that are most striking. Like *Coryphella browni*, its favourite food is oaten pipe hydroids and the two species of sea slug are often found together. [Up to 5 cm long]

Violet sea slug - *Flabellina pedata*

This is a small nudibranch but one that has fabulous colouration. The body is an overall pink-purple and the projections along its back have bright white markings at their tips. The eggs are laid as a thin white thread that is sometimes easier to spot than the sea slugs themselves because, despite their bright colours, they can be overlooked when amongst reddish seaweeds. The species is widespread, being found all around Britain and on most coasts between Norway and the Mediterranean. [Usually up to 2 cm long, but occasionally larger]

Eubranchus tricolor

The body of this nudibranch sea slug is an overall translucent white, with no markings, while the projections (cerata) on its back are quite distinctive. They are swollen in shape and have characteristic yellow bands, surrounded by white, near their tips. The dark digestive gland running up each projection is clear to see. *Eubranchus tricolor* feeds on hydroids and is often found on colonies of *Nemertesia*. This individual appeared to be the one that was photographed on the same colony of *Nemertesia ramosa* shown on page 76; the photographs were taken two days apart so it presumably liked the spot. [Up to 5 cm long]

Aeolidiella glauca

This species belongs to a sub-group of nudibranch sea slugs who specialise in preying on sea anemones. One of its favourite foods is the anemone *Sagartiogeton undatus* (page 51) which is very common in the sandy shallows of the Helford Estuary in Cornwall, where this photograph was taken. Distinguishing features of *Aeolidiella glauca* include the pale fawn pigment that forms a rim around the edge of its foot and is scattered on its back, especially all over the cerata (projections). There is also pigment on the outer half of the head and mouth tentacles. [Up to 5 cm long]

Common (or edible) mussel - *Mytilus edulis*

Close-up of mussel bed, showing the animals' siphons

The common mussel's curved, two-halved shell in blue, black and brown is instantly recognisable. Mussels normally live in large aggregations on the shore or in shallow water; the photograph opposite shows such a mussel bed in Loch Creran, west Scotland. They anchor themselves to the rocks and each other with sticky threads called byssus. These threads are planted, like the guy ropes of a tent, by the highly extendable foot which can reach out some distance from the shell (see smaller photograph on this page). Like all bivalve molluscs, mussels filter suspended food out of the water in which they live. Water is pumped in via the frilly-edged opening, and leaves by the smooth-edged outflow tube; these siphons are both visible when the shells are gaping. Despite their tough shells, mussels have a large number of predators and may be pulled open by starfish, broken open by crabs or drilled into by dog-whelks. They can even be attacked from inside by tiny pea crabs who ride in on their feeding current and then stay there to live in a permanent larder. Mussels only offer passive resistance to most predators but they can take an active stance against dog-whelks, sometimes tying them down until they starve to death (see pages 153-154). [Shell up to 10 cm long but usually much smaller]

Mussel on right has foot extended (aquarium photograph)

Common (or edible) mussel - *Mytilus edulis*

A mussel bed is also a home for other animals such as gobies, barnacles, crabs, periwinkles and dog-whelks

Horse mussel - *Modiolus modiolus*

The horse mussel is usually larger than the common mussel and is a lot more heavily built. Its bulky appearance is given perspective by the helpfully positioned crab (carrying eggs) in the photograph. The narrower end of the horse mussel's shell is also very rounded, in contrast to the more pointed end of the common mussel, but this feature may well be hidden amongst the stones or mud where it lives. Only the broad end of the shell with the openings for water entry and exit are then visible. Horse mussels can occur in dense aggregations where, aided by their byssus (anchoring) threads, they stabilise and bulk up the seabed so that a type of reef is formed. Numerous other types of animal attach themselves to the horse mussels' shell surfaces (hydroids, bryozoans, sea squirts) or hide in the crevices between the shells (worms, crabs, brittle stars) making an extremely rich community. [Shell up to 20 cm long]

Gaping file shell (or flame shell) - *Limaria hians*

This most striking of bivalves usually remains hidden inside a "nest" which is made of gravel and shell debris bound up with the file shell's own byssus (anchoring) threads. Together, many nests can form a file shell reef, a firm base where other sorts of animals can attach and make their home. The individual pictured here had been captured by a velvet swimming crab who was presumably going to try and break it open at the earliest opportunity. The crab might have been deterred, however, by the acid secretions that the file shell can produce. The attractive fringe of red/orange tentacles is always on display in this species, surrounding the oval shell which is decorated with delicate ridges. It can swim actively when disturbed, using a jet of water expelled from its snapping shells (as in the scallops); this is assisted by rowing movements of the file shell's long tentacles. [Shell up to 3 cm long]

Common European (or flat) oyster - *Ostrea edulis*

For quite a large animal, the oyster can be surprisingly inconspicuous. The approximately circular shell is rough and uneven, and this tends to make it blend in with its surroundings. The shell's lower half is convex (and may be cemented to the seabed) while its upper half is flat. This species is widely distributed but much less abundant than it used to be, with very few extensive beds remaining. Historic overfishing caused a huge decrease in numbers, and they are also susceptible to pollution and competitors such as the slipper limpet (page 151). Because most oyster farms use the Pacific oyster (see below), *Ostrea edulis* is often called the "native oyster". [Shell up to 11 cm across]

Pacific oyster - *Crassostrea gigas*

This oyster's shell is less rounded and more elongated than the European oyster's and has a very rough surface. The convex lower half of the shell is very deep and it is sometimes called the "cup oyster". The extremely wavy edge of the shell (very obvious on the two oysters in this photograph) is probably the most distinctive feature. Native to Japan and the Pacific coast of Asia, this species was introduced into Britain and other countries for cultivation, for which its fast growth makes it popular. It is now often found outside the farms and is known to be breeding wild in a number of places. [Shell up to 25 cm long]

Great (or king) scallop - *Pecten maximus*

Typical view of great scallop partly buried

Like the oysters', but unlike those of the mussel and many other bivalve molluscs, the great scallop's two shell halves differ markedly in shape. The lower half is curved like a bowl, while the upper is flat, like a lid. Both halves bear distinctive radiating ribs. Scallops use a rocking motion to make a hollow for themselves in sand or gravel, where they sit with shells gaping as they filter feed. The main photograph shows a typical partly-buried great scallop, while the smaller photograph shows one that had left its hollow, possibly while evading a predator. Scallops swim rather comically by rapidly snapping their shells, the soft tissues guiding the resulting water jets to give some control over direction. Tiring quickly, the swimming range is small, though quite enough to escape a predatory starfish. When the shells are gaping open, the attractive patterning on the soft tissue "curtain" can be seen, along with the numerous small tentacles and tiny eyes with their metallic blue sheen. Commercial dredging for scallops can devastate large areas of seabed and damage or kill many other animals in addition to catching the scallops. [Shell up to 15 cm across]

Queen scallop - *Aequipecten opercularis*

Queen scallop with covering of sponge

Queen scallop swimming

This bivalve is similar to the great (or king) scallop but, as its name suggests, tends to be smaller. A better distinguishing feature is that the upper half of its shell is curved like an upside down bowl, rather than flat like the great scallop's. The upper half is actually slightly more cupped than the lower. The queen scallop is a more vigorous swimmer than its larger relative but uses the same snapping motion. It will often swim several metres at a fair pace when disturbed or threatened by a predator (see small photograph). Growths of sponge are found on many queen scallop shells (see main photograph) and this arrangement is thought to benefit both parties. The sponge is less likely to be grazed by a predator, such as the sea lemon sea slug (page 170), than if it was on a rock; and the sponge will help to keep starfish off the scallop. [Shell up to 10 cm across]

Common cuttlefish - *Sepia officinalis*

Cuttlefish hunting; the jet-propulsion siphon can be seen protruding beneath its head

I have spent more fascinating hours watching cuttlefish than any other marine animals. Encounters with them are almost always memorable and you are often left with the impression that you have been observed as much as observing! Apart from sea mammals, cuttlefish (along with their octopus relatives) are the most **intelligent creatures** we meet in the sea and have a huge range of intriguing aspects to their behaviour; these seven pages are intended to introduce some of those aspects. Other species of large cuttlefish are rare in British waters and, while very small individuals could be confused with the little cuttle (*Sepiola*, see page 187), the common cuttlefish has a highly distinctive appearance. The broad and slightly flattened body, up to 30 cm long, is fringed by a fin on each side which runs from just behind the head right back to the rear. The mouth is surrounded by eight arms, two of which are sometimes raised above the head when the cuttlefish is approached. In addition to the eight arms, there are also two much longer extendable tentacles which can shoot out to catch prey such as fish or prawns. The arms, along with the razor-sharp beak hidden behind them, can deal swiftly with

Common cuttlefish - *Sepia officinalis*

This cuttlefish has the pattern shown by many when they are first encountered

crabs (a favourite food) after they are grasped from behind and smothered to neutralise their threatening claws. If startled when met, a cuttlefish may use its full **escape mechanism**, where the body assumes a shape for maximum streamlining and a powerful burst of jet propulsion thrusts it rapidly backwards. At the same time, it can release a cloud of black ink which will hang in the water and momentarily distract a pursuer while the cuttlefish escapes. If still pursued, larger clouds of ink are poured out. Once used by artists, the ink was called sepia, as in the animal's scientific name. If not startled when first encountered, the cuttlefish may well move slowly away, using a little gentle jet propulsion aided and stabilised by the rippling motion of its fringing fins. It is then that their most impressive skill, the total **control of colour and patterning**, can be appreciated. They may be less famous for it, but cuttlefish are much more skilful colour-change artists than even chameleons. If a cuttlefish swims off over a varying seabed, its shade can change instantly to match its surroundings, going dark over kelp-covered rocks and almost white over sand. The animal seems to calculate all the angles, apparently matching the seabed against which the observer is seeing it, rather than that directly below it. If it comes to rest on the bottom, it will blend in perfectly. The top photograph on page 182 shows a cuttlefish as

Common cuttlefish - *Sepia officinalis*

it first comes to sit on a gravel seabed; a few seconds later (middle photograph) it has matched the gravel perfectly. To further improve **camouflage**, projections on the skin's surface will change its texture to replicate that of the surroundings. Cuttlefish will also flick sand or gravel up over themselves and gradually sink into the seabed, leaving only the tops of their heads and backs exposed. When in such a position, only the most sharp-eyed diver will spot them. In addition to simply matching their surroundings, cuttlefish use all sorts of patterns in an attempt to break up their outline and generally confuse potential predators. From hatching, young cuttlefish can display at least thirteen types of **body pattern**, made up from over thirty different components. Whatever the exact purpose of them all, and even the scientists who have logged them remain unsure of this, watching a cuttlefish flash through part of its repertoire is an awe-inspiring sight. One of the most common patterns is known as the "white square" (bottom photograph) which is accompanied by a white stripe on the head and a small white triangle at the rear.

This cuttlefish has just settled on a gravel seabed

A few seconds later, it has matched its background

Another individual, showing the white square pattern

Common cuttlefish - *Sepia officinalis*

Times when particularly impressive fast pattern changes can be observed include when cuttlefish are hunting (see three photographs on this page) or when they seem to be unsettled by the approach of a large fish (top pair of photographs on page 184). They sometimes respond in this unsettled way with divers but, at other times, appear unperturbed. The secret of **rapid pattern change** is explained by

This cuttlefish changed from its resting pattern (top photograph) as it extended its tentacles to catch a two-spot goby (lower photograph, see right-hand of picture). Such colour changes may fatally distract prey from the rapidly approaching tentacles

Another individual nearby turned virtually white as it grabbed its prey; the doomed goby can just be seen caught by the suckers on the ends of the extended tentacles

Common cuttlefish - *Sepia officinalis*

This resting cuttlefish (top photograph) changed colour and pattern dramatically when a large ballan wrasse (at left of lower photograph) came close

the presence of special cells, known as chromatophores, within cuttlefish skin. These are effectively little flexible bags of pigment which, when expanded by muscular action, make the area of skin appear dark. When the bags are allowed to contract under the power of their own elasticity, the skin appears pale. In addition to its other roles, pattern control is also used in **courtship**. Male cuttlefish display brilliant zebra stripes at courting time in order to impress females and warn off competitors (see main photograph opposite). **Mating** is achieved by the male grasping the female (see photograph below) and passing her a packet of sperm, using one of his arms specially adapted for the purpose. After mating, a male will often defend the female while she lays her eggs (see small photograph opposite), though there can be a delay of several days. **Clumps of eggs**, dyed black with ink and known as "sea grapes", can be seen attached to seaweed or eel-grass in the summer months (top photograph,

Cuttlefish pair mating; the male is on the right

Common cuttlefish - *Sepia officinalis*

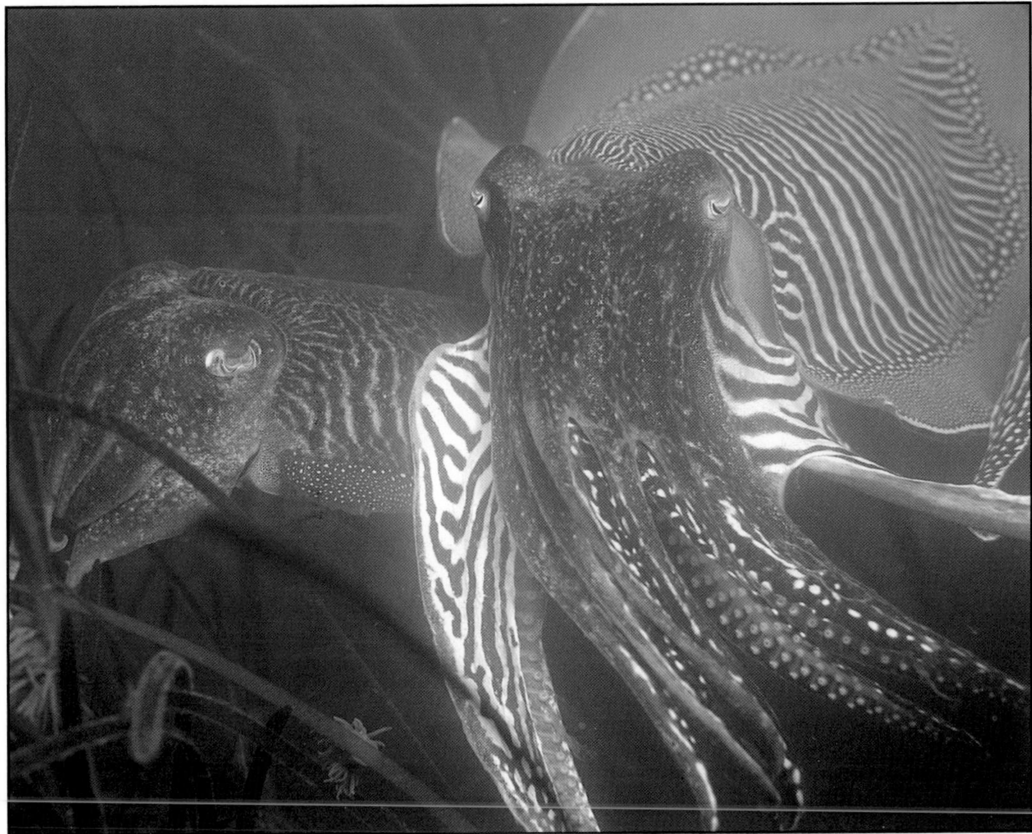

A courting couple, the male with dramatic stripes

page 186). The **eggs hatch** after two to three months and miniature cuttlefish, around finger-nail size, emerge (middle photograph, page 186). As mentioned earlier, the junior cuttlefish immediately show good powers of patterning control. Even so, they are very vulnerable to predators, and mortality at this stage in their life is high. Females only breed once and die soon after laying their eggs. Unlike their relative the octopus, they do not take care of them. After cuttlefish

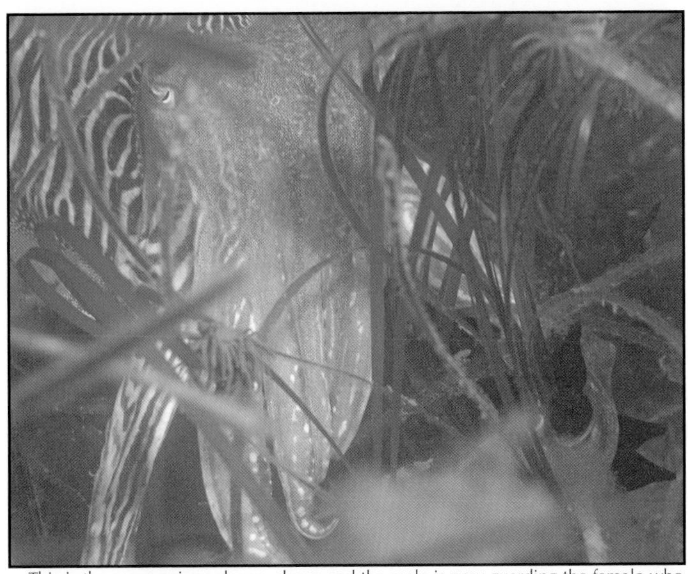

This is the same pair as shown above and the male is now guarding the female who is laying eggs. She is largely hidden by the male and the eel-grass, but her brown tentacle can be seen wrapped around the eel-grass as she attaches the eggs

Common cuttlefish - *Sepia officinalis*

have died and decomposed, their internal skeleton or **cuttlebone** (bottom photograph, this page) is often found washed up on beaches. In life, this is the animal's buoyancy organ and it consists of stacks of thin-walled chambers that can be filled with either liquid or gas in order to give precisely the right degree of lift. Cuttlefish spend the winter in relatively deep water, such as that in the western English Channel, though they cannot inhabit very deep water because the cuttlebone could implode under high pressure. They move into shallow coastal waters to breed in the spring and summer. It is then that divers can see them in large numbers, and snorkellers may even see them close to the shore. Numbers of cuttlefish found inshore at particular locations fluctuate markedly, with large numbers being seen in some years and hardly any in others. The effect of the **increasing fishing pressure** on cuttlefish is a serious concern, especially when large numbers of breeding individuals and their eggs are destroyed. Anyone who has met these fabulous animals at close quarters will be desperate to see them continue to thrive in our waters. [Up to 30 cm long]

Cuttlefish eggs

A new hatchling

Cuttlebone

Little cuttle - *Sepiola atlantica*

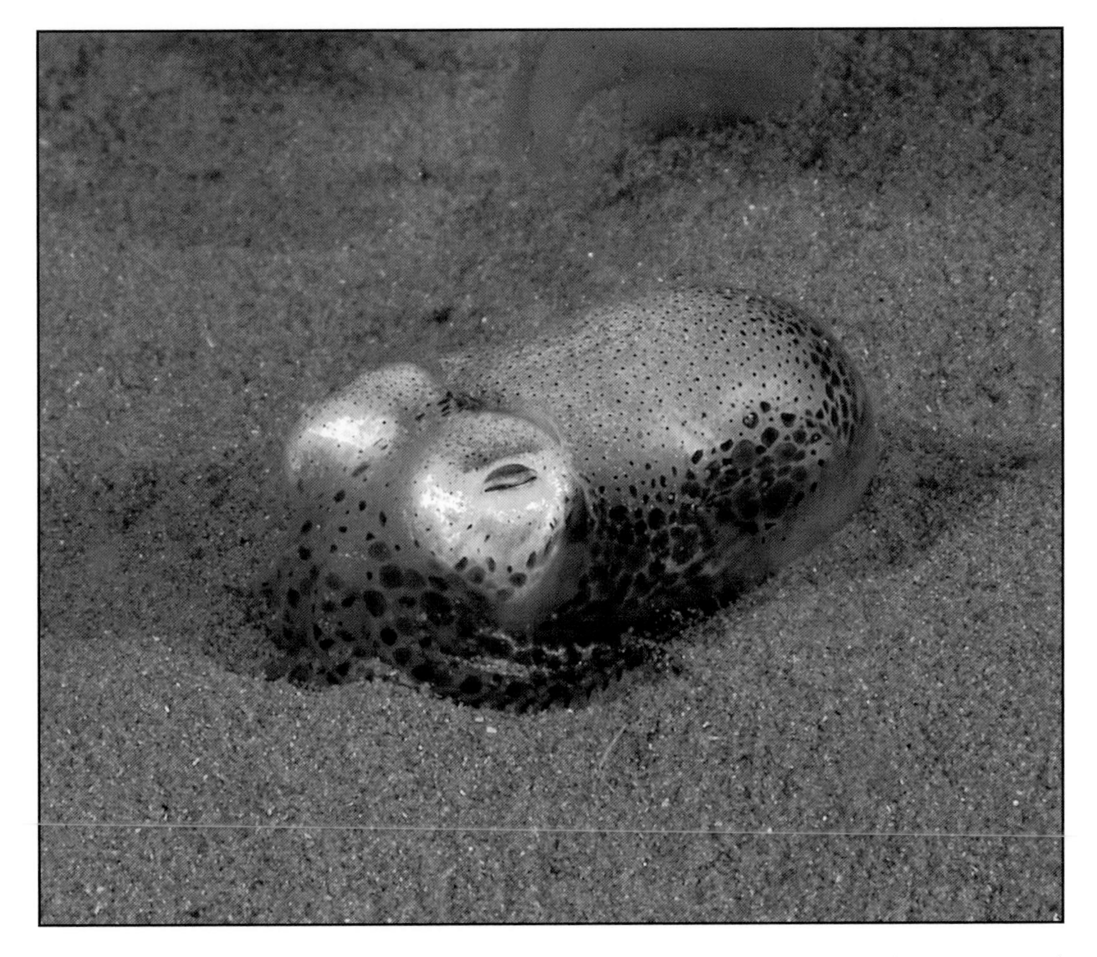

The little cuttle, *Sepiola*, has the appearance of a miniature *Sepia*. It is very difficult to spot underwater and only close inspection will reveal that it is a rather different shape to its larger relative. The body is cup-shaped and there is a pair of lobe-like fins that do not run the entire length of the body. Other characteristic features include its very protrusive eyes and the greenish tinge to the tops of the eye bulges. The little cuttle's buoyancy control system is less sophisticated than that of the common cuttlefish but it can hover effectively or jet away quickly. The tiny clouds of ink released by this species when disturbed appear quite comical to the diver, but can still be used effectively in combination with colour control. By discharging a dark cloud of ink at the same moment that they jet away, and by turning their body from dark to pale, a predator's attention can be distracted to follow the ink cloud and the cuttle can escape unnoticed. The photograph above shows the role of individual colour cells (chromatophores) in cuttlefish pattern control. Around the bottom of the body and on the tentacles, the cells are expanded forming dark discs. On the back, they are contracted into small spots so the skin appears mainly pale. Little cuttles live on sandy seabeds and spend much of the time partly buried with only their eyes exposed, watching for predators or hapless small crustaceans which they will capture on emerging from the sand. [Up to 5 cm long]

Lesser (or curled) octopus - *Eledone cirrhosa*

The octopus may be one of those animals usually associated with warmer climes but this species can be a regular sighting for observant divers in many areas of Britain. Like all octopods, it has a **soft bag-like body** which contains no skeletal support (unlike its cuttlefish relatives which have the cuttlebone) and eight long sucker-bearing arms. That body can be moulded into virtually any shape and squeezed through the tiniest crevice in an attempt to catch prey or evade a predator such as a conger eel, cod or sea mammal. An octopus will often remain hidden in a hole or crevice during the day. The smaller photograph on the opposite page shows an octopus at the entrance to its home, with an eye and part of its body visible; an arm is spread out to partially cover the entrance. If an octopus itself cannot be seen, empty crab carcasses near the lair may give its presence away. Crabs, which form a large part of their **diet**, are enveloped in the arms of the octopus, pierced by their sharp jaws and killed by injected poison. Digestive enzymes are pumped in and the resulting crab "soup" is sucked out. The larger photograph on the opposite page shows an octopus just finishing off its meal of an edible crab and discarding the pieces of empty shell. Although they don't simply shed eggs and sperm into the seawater like many marine animals, the **coupling** of octopods does not appear to be an intimate affair. The timid lovers seem to keep as far apart as possible and the male caresses the female with an outstretched

Lesser (or curled) octopus - *Eledone cirrhosa*

An octopus finishes off its meal of an edible crab and discards the remains

Octopus hiding in crevice, small spider crabs lurk nearby

arm before inserting packages of sperm into her egg ducts. One of the male's arms is specially adapted for this purpose. The female octopus lays her eggs in large bunches within rock crevices or beneath stones. She stays to guard them until they hatch, hosing them with water from her funnel to keep them clean and aerated. The **devoted mother** barely feeds during this vigil and may well die soon after it is completed. Like cuttlefish, octopus are extremely intelligent. When kept in aquaria, they can learn their way around mazes in order to get food. A distinguishing feature of the lesser (or curled) octopus is that there is only a single row of suckers on each arm. The larger common octopus (*Octopus vulgaris*) has a double row of suckers and is found occasionally on the south coast of England; this representing the northern extent of its range. [Lesser octopus arm span up to 70 cm; common octopus arm span up to 3 m]

Chapter 7
BRYOZOANS

Sea mosses
A number of marine animals resemble plants but bryozoans take this to the extreme, with the pale brown growths of many species looking just like poorly seaweed at first glance. Bryozoa means "moss animal" and members of this phylum are sometimes called "sea mosses".

Hairy tentacles
Bryozoans are stationary colonies of tiny animals called zooids. Most of these zooids are specialised for feeding and each has a ring of tentacles which surrounds its mouth, so it looks a little like a hydroid or miniature sea anemone. However there are no stinging cells (as there are on hydroid and sea anemone tentacles) and the bryozoans' tentacles are instead equipped with numerous hairs called cilia. These cilia beat rapidly and produce currents which carry microscopic organisms, such as bacteria and single-celled algae, into the bryozoan's mouth. There is a U-shaped digestive tract to process this food, and any undigested remains pass out of the anus which is positioned just outside the ring of tentacles.

Living in a box
Each individual zooid lives in its own case, less than 1 mm across, which is usually shaped like a box. The walls of this box are often reinforced with calcium compounds. Numerous boxes joined together in the colony produce the mesh-like appearance that is so characteristic of bryozoans. The tentacles which protrude from each box can be seen when there is active feeding underway, but they are withdrawn instantly at the slightest disturbance.

Specialisation and reproduction
Within most bryozoan colonies, there are zooids with specialisms other than feeding. Some attach their colony to the seabed, while others defend the colony against the larvae of other creatures that may settle on it or against larger animals (such as worms) that may crawl over it and cause damage. There are also zooids that specialise in reproduction. All bryozoan colonies are hermaphroditic (male and female at the same time) but some species have separate male and female zooids within a colony, while others have zooids that produce both eggs and sperm. Fertilised eggs are usually brooded within the colony and the resulting larvae swim for a few hours before finding a suitable place in which to settle. Each larva can transform into a zooid that then buds (reproduces asexually) to produce more zooids and form a large colony.

Sea-mat - *Membranipora membranacea*

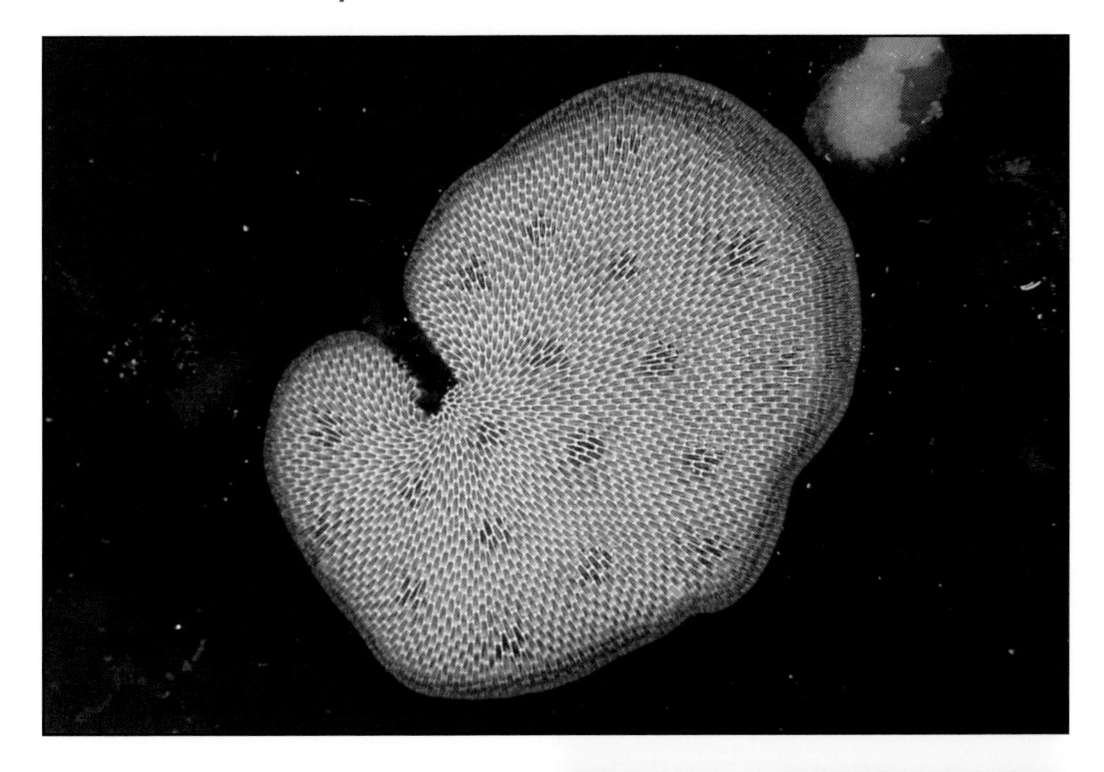

The distinctive colonies of the sea-mat are seen as a layer of fine lacy mesh on fronds of seaweed, particularly kelp. Though tiny, the individual rectangular "boxes" that make up the colony are clearly visible. Arranged in rows like brickwork, they give the colony enough flexibility to cope with bending of the kelp frond. The main photograph shows a colony just a few cm across, but they can cover much larger areas and spread rapidly by budding. By growing near the base of the frond, a colony ensures that it will not all be lost if the kelp frond becomes damaged and loses material from its outer edges. Several species of sea slugs (such as *Onchidoris muricata* shown in the bottom photograph, see also page 167) feed specifically on this bryozoan. Such mesh-encrusted fronds in shallow water are a good place to find a variety of sea slugs and their eggs. ["Boxes" less than a mm across but colonies can cover large areas of a kelp frond]

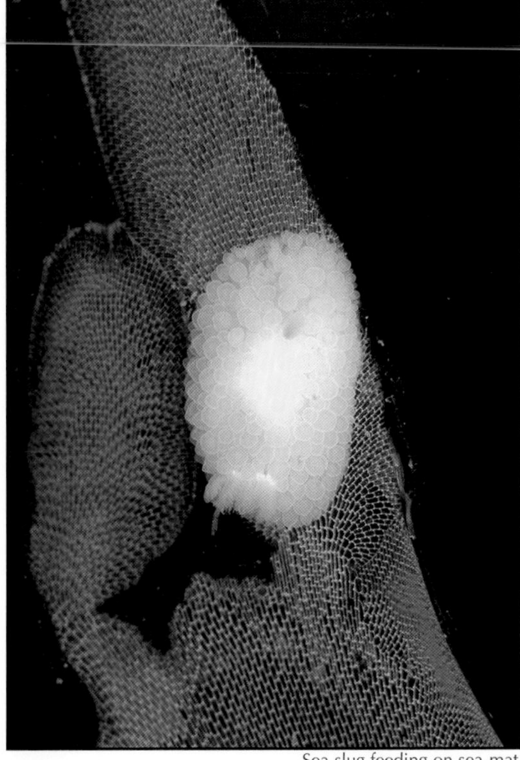

Sea slug feeding on sea-mat

Sea chervil (finger bryozoan) and hornwrack growing in the Menai Straits

Sea chervil - *Alcyonidium diaphanum*

Sea chervil (also called the "finger bryozoan") is easy to recognise although its fleshy texture and finger-like growths make it look more like a sponge than a typical bryozoan. Usually brownish in colour and with wavy surfaces, the fingers can be branched and are often found in bunches. The slightest amount of tide or current tends to push the bunches almost flat to the seabed on which they are attached. The sea slug, *Acanthodoris pilosa,* eats this species of bryozoan and lays its white egg ribbons on the fingers (see slug and eggs in the small photograph above, also the slug on page 168). Sea chervil is said to cause a type of skin irritation known as "Dogger Bank itch" by fishermen in the North Sea. [Fingers up to 50 cm long]

Hornwrack - *Flustra foliacea*

The frond-like colonies of hornwrack can cover quite large areas of rocky seabed in current-swept areas although their drab beige colour means they might not grab your attention. Pieces are often washed up on beaches where they look just like pale dried seaweed; but the telltale bryozoan "mesh" pattern can be made out on both dead pieces and on live growths underwater. Colonies only grow for part of the year so annual growth lines can sometimes be seen. This species is said to produce a characteristic lemon smell but this is not obvious in dried specimens or from inside a diving mask! The small photograph above was taken beneath the famous arch of Cathedral Rock at St Abbs. [Colonies up to 15 cm tall]

Spiral bryozoan - *Bugula plumosa*

A number of bryozoan species, including several others which belong to the *Bugula* genus, occur as buff coloured colonies that can form a "turf" on rocks or other hard surfaces. These growths are easily overlooked but close inspection reveals that many of them have surprisingly attractive forms. The tuft-like colonies of *Bugula plumosa* are shaped rather like tiny Christmas trees, but with their branches arranged in a spiral. Colonies are often found in small groups and, like other bryozoans, are preyed upon by certain species of sea slugs. [Up to 8 cm tall]

Potato crisp bryozoan (Ross or rose "coral") - *Pentapora foliacea*

This distinctive species is not a coral of course, so the potato crisp bryozoan name is less misleading. However, it forms open mounds made up of stiff interlocking "leaves" so the overall effect is that of a coral head or a cabbage. A close look at the "leaves" reveals the typical fine mesh-like structure of a bryozoan. Colour can range from dark orange-brown to sandy buff, but part of the mound may be overgrown by algae or hydroids. Many small animals can reside in the cavities formed between the "leaves". This bryozoan is most common in the west and south west of Britain. [Mounds up to 50 cm across]

Chapter 8
ECHINODERMS
Starfish, brittle stars, sea urchins, sea cucumbers & feather stars

The name, echinoderm, means "spiny-skinned" and refers to the skeleton of these animals, which is composed of bony plates embedded in the body wall. These plates may protrude to form spines and take on very different forms in the various sub-groups. They are relatively loosely arranged in starfish for instance, but form a rigid structure in sea urchins. Echinoderms have no front or back end because they are based on what is called radial symmetry, rather than the bilateral symmetry of many familiar animals. Starfish, for example, are just as likely to lead with one arm as with any other, and have no need to turn their body when changing direction. Some urchins and sea cucumbers operate as though they have a front end, but their basic body plan is still based on radial symmetry. Most echinoderms have separate sexes (rather than being hermaphrodites) and release their eggs and sperm into the water, so that fertilisation is external. A planktonic larval phase usually follows. The diagram below shows a generalised starfish life cycle followed by several (but not all) common species. After spawning, a fertilised egg develops through various larval stages; the final one of which houses a miniature starfish. This stage attaches to the seabed and releases the young starfish that then grows to maturity.

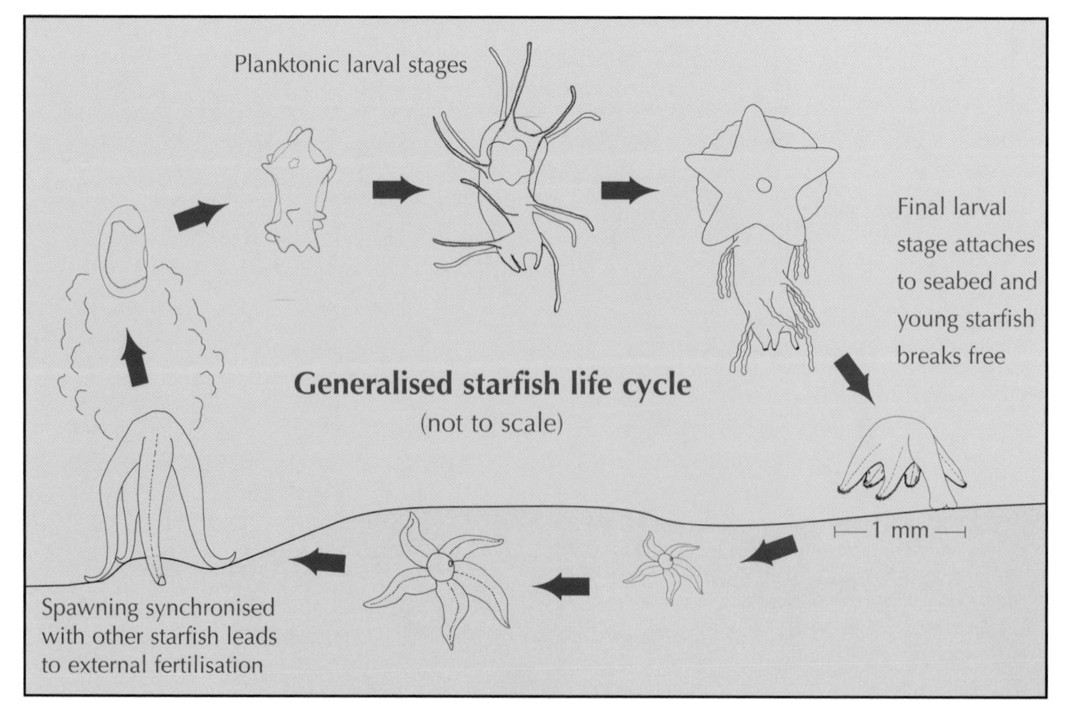

Starfish close-up, showing tube feet, spines and pedicellariae

Starfish features

The first thing you notice if you turn a starfish over is the huge number of tiny extendable legs, known as tube feet, bursting from the groove underneath each arm. All echinoderms have tube feet, operated by a special hydraulic system, even though they may be less conspicuous than those of starfish. Starfish tube feet are usually equipped with small suckers for gripping onto firm surfaces, and are used for walking the animal along and, in some species, for pulling open prey (see pages 197-198). The close-up photograph above shows the tube feet underneath the arm of a spiny starfish. It also shows the upper side of the arm, on which the prominent spines are surrounded by bunches of tiny structures, called pedicellariae, that have an intriguing function. Slow moving animals with a hard exterior can present an ideal home for encrusting animals, such as barnacles, looking for a free ride. The pedicellariae are like miniature pairs of jaws which, when touched by a small animal looking for somewhere to settle, will grasp it and kill it. Starfish have amazing regenerative powers. A single detached arm can produce a whole new starfish, as long as part of the central body is also present. This explains the bizarre shape of some individuals (see page 199).

Brittle stars

These animals look like emaciated starfish but have a different body layout because their arms are very clearly demarcated from the central disc. The bony plates of the arms form a continuous articulated armour which gives the whole arm the appearance of having joints. The arms are easily broken, hence the name, and can be jettisoned to permit escape from a predator, and then regenerated. Brittle stars move much more rapidly than starfish, using snake-like movements of their arms to propel them across the seabed. The tube feet simply give the arm some grip, rather than walking the animal along.

Various echinoderms on a gravel slope near Oban; common starfish, common sunstar, common sea urchin and feather stars

Sea urchins

The bony plates possessed by sea urchins are more imposing than those of any other echinoderms and fit together to form a solid structure, known as a test. The usually impressive spines are mobile, being attached to the test with ball-and-socket joints, and they can help the tube feet with movement or be locked rigidly in position to anchor the urchin in a rock crevice. The tube feet are similar to those of starfish, though they are much longer in order to reach out beyond the spines.

Sea cucumbers

These cucumber-shaped animals can be visualised as starfish with no arms that have grown extremely tall and thin. Some sea cucumbers, like the cotton-spinner, crawl over the bottom eating organic deposits. Others wedge themselves into rock crevices or burrow into soft sediments. From here, they hold out feeding tentacles, which are highly modified tube feet that surround their mouths, to catch suspended food. More conventional tube feet along the rest of the body help the cucumber with walking or burrowing. The skeleton of sea cucumbers is simply a few scattered bony elements in their leathery skin. As with other echinoderms however, the soft elements of the body wall can alter consistency, being soft when the cucumber needs to get through a narrow crack and then stiff when it needs to wedge itself in position.

Feather stars

The aptly named feather stars have a long fossil history and are thought to be the most primitive echinoderms. The typical feather star features are described on page 227.

Common starfish - *Asterias rubens*

The most familiar starfish to non-divers, it can be found on the lower shore and at virtually any depth, on sandy and stony seabeds as well as on rocks. It is usually orange or pale yellow/brown, sometimes red or purple, with a cream underside. The five (occasionally four or six) arms taper down towards their tips. The starfish's upper surface is covered with small blunt spines, with a marked line of larger spines down the middle of each arm. Belying their sluggish image, these starfish can be **fearsome predators** of bivalve molluscs such as mussels and clams. The top photograph on page 198 shows a starfish in characteristic feeding pose arched over a clam (the stripes on its shell are just discernible). By using this position, it puts the maximum number of tube feet to work on pulling open the mollusc, and also positions its mouth right next to the crack between the shell halves. It is not clear whether a steady pull or a series of tugs is used, but the force applied by the tube feet has been estimated as being equivalent to 5 kg, and the hapless mollusc eventually tires. As soon as the tiniest crack appears between the shell halves, only a fraction of a millimetre across, the starfish pushes its stomach out through its mouth and slips it through this crack into the mollusc's interior. Digestive juices get to work and the battle is over. Common starfish can congregate in **large aggregations** where food is plentiful and can be serious pests to mussel or oyster fisheries. The bottom photograph on page 198 shows a

Common starfish - *Asterias rubens*

Feeding on a clam

A mussel bed under attack

Common starfish - *Asterias rubens*

Regenerating from a single arm and part of the central body

A purple individual, digging down in search of buried prey

Common starfish - *Asterias rubens*

Preparing to spawn

Spawning

Common starfish - *Asterias rubens*

Righting itself

mussel bed near St Abbs (south east Scotland) which was under heavy attack. Bright silvery spots among the mussels mark shells emptied by starfish. Past attempts by shellfish collectors to kill dredged up starfish by simply chopping them in half and throwing them back into the sea were doomed to failure, because each half starfish simply grew another half and carried on as before! The top photograph on page 199 shows the "comet" shape that results when a starfish is **regenerating** from a single arm and part of the central body. As well as preying on animals found on the surface of the seabed, common starfish often seem to dig part of themselves down into sand or mud in search of buried bivalves (see bottom photograph of a purple individual on page 199). As in most starfish, the sexes of this species are separate. **Fertilisation** takes place when females and males shed their eggs and sperm (respectively) into the seawater at the same time. A chemical released with the eggs is thought to inspire both males and females in the vicinity to start **spawning**. The top photograph opposite shows a starfish preparing to spawn, raised up onto its arm tips on a rocky ledge. Prominent locations are often sought, with the top of kelp fronds a favourite, and the bottom photograph opposite shows an individual in the midst of spawning while in such a position. The milky cloud can be seen drifting off in the current. The photograph on this page shows how a starfish can right itself by twisting its arms and pulling itself over with the versatile tube feet. [Up to 50 cm across but rarely more than 30 cm and usually smaller]

Spiny starfish - *Marthasterias glacialis*

Spiny starfish spawning

With its large body covered in conspicuous knobs and spines, the spiny starfish can be an impressive animal. Mainly restricted to the western side of Britain, it can be very common in rocky or sandy and muddy habitats and over a large depth range (see top pair of photographs opposite). Typically pale blue, they can also be grey, brown or white. The purple arm tips are most noticeable on pale individuals. Although the arms of this species appear heavily armoured with three rows of large spines running their length, they are quite soft and will be readily jettisoned if hurt or over-handled. Fortunately, they can be regenerated and spiny starfish with a mixture of normal arms and smaller ones in the process of being re-grown are a common sight. The tiny pincer-like organs for removing settlers (pedicellariae) are particularly prominent in this species and form obvious cushion-like wreaths around the large spines (see close-up photograph on page 195). Spiny starfish are voracious predators and, in addition to preying on bivalve molluscs, eat crustaceans, fish and other echinoderms such as starfish and sea urchins, dead or alive. Like the common starfish, this species can be found spawning, often crawling onto the top of kelp to do so (see above and bottom pair of photographs opposite). [Up to 80 cm across but usually much smaller]

Spiny starfish - *Marthasterias glacialis*

Spiny starfish are common from shallow seaweed meadows…

…to deep reefs

Two spawning starfish

This huge starfish found spawning was 80 cm across

Northern starfish - *Leptasterias muelleri*

This species looks rather like a small version of the common or spiny starfish (pages 197-203). It is, however, more spiny than the common starfish and lacks the obvious wreaths around the spines possessed by the spiny starfish. A clear line of spines can be seen down each arm in this young starfish. In older individuals, the spines are more randomly scattered. The northern starfish is usually purplish, becoming paler near the tips of the arms, but individuals found on the shore or in shallow water may be green. This is due to algae living in their tissues. As the name suggests, it is found mainly around the north of Britain. This photograph was taken off Anglesey where it appeared to be quite common. [Usually no more than 10 cm across, often smaller]

Bloody Henry - *Henricia oculata*

Common in wave or current swept rocky areas on southern and western coasts, the bloody Henry has an upper side that is usually coloured a dramatic pink, purple or crimson, and which has a rough sandpaper-like texture. The underside is a pale sandy colour. It has quite a different appearance from the common, spiny and northern starfish, because its five arms are round in cross-section, lack any obvious spines and look stiffer than those of the other species. It is a suspension feeder for at least some of the time, raising its arms up to catch food particles, and also browses on sponges. It has a smaller stomach than the more predatory starfish. An almost identical species, *Henricia sanguinolenta*, has a more northern distribution so the two species probably overlap on the west coast of Scotland. [Up to 20 cm across]

Common sunstar - *Crossaster papposus*

Though they possess many more arms than "standard" starfish, sunstars are very similar to them in basic form. The very striking common sunstar can have anything from 8 to 14 arms, but 10 to 12 is the most usual. The upper surface of the body and arms is very spiny and there are distinctive brush-like spines around the edges of the arms which create a rather "jagged" appearance. Colouration can be dull brown, but it is usually brilliant red or orange with pale bands on the arms that produce an impression of concentric rings like a target. The small white spot, often visible off-centre on the body, is the sieve plate (madreporite) which is the opening into the hydraulic system.

This species is a voracious predator, eating a wide variety of animals. Other echinoderms such as starfish, brittle stars and sea cucumbers are a significant part of the diet (see photographs on page 206). The sunstar in the top photograph on page 207 is spawning and the individual orange eggs being released are visible. These large and yolky eggs are unlike the eggs of many starfish and they develop without a feeding larva stage. The common sunstar can be found all around Britain but is rarely seen on the south coast. The photograph above also shows daisy anemones (page 53) and light-bulb sea squirts (page 231). [Up to 35 cm across]

Common sunstar - *Crossaster papposus*

As this sunstar crawls across the seabed, the brittle stars have wisely moved away to give the predator its "own space"

A brief look under this sunstar showed its stomach lobes had enveloped one common starfish and were starting on another

Common sunstar - *Crossaster papposus*

Spawning sunstar, with the individual orange eggs visible

The tyre shows the impressive size of this sunstar. Inset: juvenile on seaweed

Purple sunstar - *Solaster endeca*

The purple sunstar is only common in the north of Britain. It has a rough upper surface but the spines are much less obvious than in the common sunstar. There are usually 9 or 10 arms, but this can vary from 7 to 13. Colour is normally a vibrant pink-purple but, despite the name, pale orange or cream-coloured individuals are also common. The paler coloured tips to the arms are often curled up, as on this large sunstar found on the silt-covered wreck of the Shuna in the Sound of Mull (main photograph). Like the common sunstar, this species is a voracious predator on other echinoderms. The two smaller photographs show a sunstar that, when turned over, was found to be eating a surprisingly large sea cucumber. After a quick look, it was turned back over and left to get on with its meal. [Up to 40 cm across]

Hunched up in feeding posture......

......a sea cucumber is being consumed

Cushion star - *Asterina gibbosa*

The relatively large central portion and short blunt arms of this cushion star give it an almost pentagonal shape. It is usually much smaller than the other starfish described here, and is mainly found on the shore or in very shallow water. The upper surface of its body is rough, covered in small bundles of short spines. These spines, and those around the edges of the arms, are usually orange but the overall colour is quite variable, ranging from green or brown through to cream. Most starfish have entirely separate sexes but members of this species start out as males and become females later in life. It is a scavenger of dead plant and animal matter and is found mainly on west and south west coasts of Britain. [Up to 5 cm across]

Red cushion star - *Porania pulvillus*

With its short arms, plump stature and velvet-like appearance, this creature really lives up to the name of cushion star. It is usually brilliant red, with scatterings of translucent small soft projections on its back, and some additional creamy-coloured markings. It occurs around western coasts of Britain but is quite rare in the south; I have mainly seen it on the west coast of Scotland. Often found on soft corals such as dead men's fingers on which it is thought to feed (see photograph on page 13). [Up to 12 cm across]

Seven armed starfish - *Luidia ciliaris*

A relatively primitive starfish, *Luidia ciliaris* has a distinctive appearance, although its orange colouration is shared by other species. The seven long arms are rather soft and only start to taper near their tips. A very prominent fringe of white spines runs round the edge of each arm. Unlike the majority of starfish, its tube feet end in tiny knobs rather than suckers (see smaller photograph). *Luidia* is found on both sandy and stony seabeds all around Britain except for south east England. It can move surprisingly quickly and feeds mainly on other echinoderms such as brittle stars, starfish and urchins. An adult can apparently produce over two hundred million eggs in a year, so mortality at the egg and planktonic larval stages must be enormous. [Up to 60 cm across]

Seven armed starfish - *Luidia ciliaris*

The seven armed starfish can move surprisingly rapidly across the seabed

This starfish is in hot pursuit of the brittle stars, which are desperately trying to avoid its grasping arms

Sand star - *Astropecten irregularis*

This burrowing species has a classic star shape with five fairly stiff arms that taper noticeably along their entire length. Each arm is fringed with long pale spines and normally has a purple spot at its tip. The overall colour is typically a pinkish sandy hue, though this may vary. With a much more flattened body than most starfish, and with pointed instead of sucker-bearing tube feet, the sand star is well adapted for burrowing into soft sediments. While burrowing, it is thought to eat shellfish, crustaceans and worms encountered along the way. Sand stars can sometimes be found moving over the surface of a sandy seabed and they may consume small fish if these are slowed down by ill health. The sequence of photographs on this page shows a sand star disappearing into the sand over a period of only about 30 seconds. Rather than sliding down edge-on into the sand, as might have been expected, it sank straight downwards. Sand stars are found in a wide range of depths around most of Britain where there is soft sand or mud into which they can burrow. The photographs opposite show a few of a large number of sand stars found spawning in Loch Kishorn. A prominent bump was obvious in the centre of the upper side of most starfish, with the bump appearing largest on the individuals that were most actively releasing spawn. Also, the larger the bump, the more tattered and depleted its owner appeared. [Up to 15 cm across]

Sand star burrowing into the seabed

Sand star - *Astropecten irregularis*

Three spawning sand stars; the one with the largest bump (bottom right) looked in a poor state with parts of its arms missing

Common brittle star - *Ophiothrix fragilis*

Common brittle stars, with arms raised for catching suspended food

Common brittle star - *Ophiothrix fragilis*

This brittle star has flexible arms covered in long glassy spines which give them a very bristly appearance; and the central body also bears spines. Colouration is very variable, with red, yellow and blue often mixed in with brown and grey. The arms almost always have alternating light and dark bands. It is usually found amongst rocks and stones, sometimes forming huge aggregations in deep water. Within these, up to 10,000 brittle stars per square metre have been recorded. The brittle stars' arms are raised up into the passing current where their tube feet filter out suspended food matter, which is then passed along the arms to the central mouth. If the current gets too strong, the brittle stars link arms to form a huge mat that is less likely to get swept away. [Up to 20 cm across]

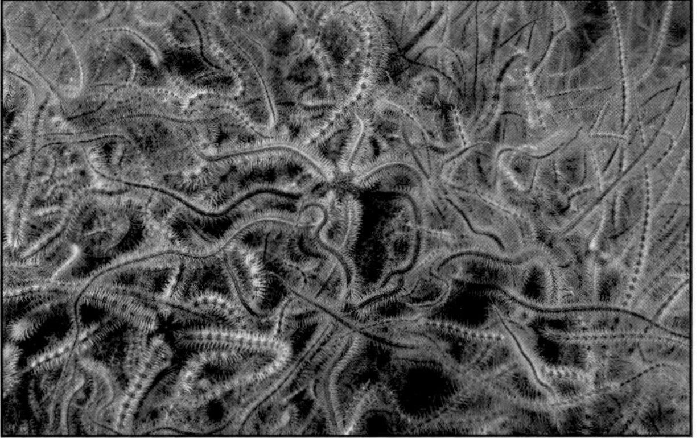

Black brittle stars on left, common brittle stars on right; also common starfish, common sea urchin and sea loch anemones

Black brittle star - *Ophiocomina nigra*

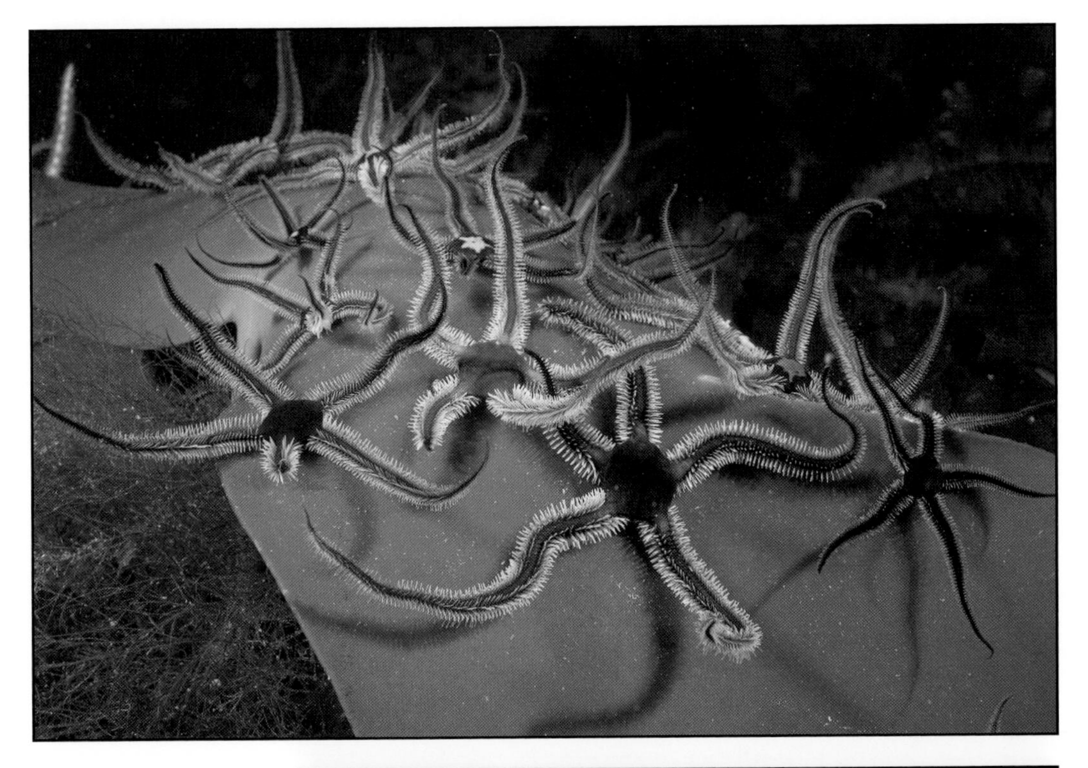

A distinctive species that appears clearly different from the common brittle star (see pages 214 to 215 and opposite). Its central disc has no spines and looks smooth rather than bristly. The spines on its arms also look more neatly arranged, like the teeth on a comb. The brittle stars can be black or almost any shade of brown, with an attractive pattern sometimes present on the disc. Though often found in aggregations, these tend to be much less dense than those of the common brittle stars, with each individual having something of a "personal space". Mixed groupings of both species are also common. Black brittle stars obtain their food by catching floating material in a net of mucus or by browsing on algae, organic deposits or carrion. [Up to 25 cm across]

Crevice brittle star - *Ophiopholis aculeata*

Crevice brittle stars (most obvious) with common and black brittle stars also present

The crevice brittle star lives up to its name and is usually found partially hidden in a rocky crevice or beneath a boulder with just the arms protruding. The arms are distinctive because they tend to be twisted up rather than lying out straight, and are usually marked with banding. The central disc is covered with a pattern of large plates, separated by granules. Overall colour is typically reddish or purplish brown. The arm banding is obvious on the individuals in this photograph; also present are common brittle stars (on the left with very prominent glassy spines) and black brittle stars (coloured black and dark brown). The crevice brittle star is much more common in the north of Britain than in the south. [Up to 15 cm across]

Sand brittle star - *Ophiura ophiura*

Normally found on sand or muddy sand, this species has a relatively large disc and shorter, stiffer arms than many brittle stars. Its arms are much less bristly than the common and black brittle stars. It is typically a drab sandy brown but there may be an attractive pattern on the central body created by the plates which cover its surface. The sand brittle star can burrow into the sediment or move quite quickly across its surface by using its arms in a "swimming" or "rowing" motion. The animal at the top right of the photograph is a burrowing anemone (see page 59). [Up to 25 cm across but usually much smaller]

Ophiura albida

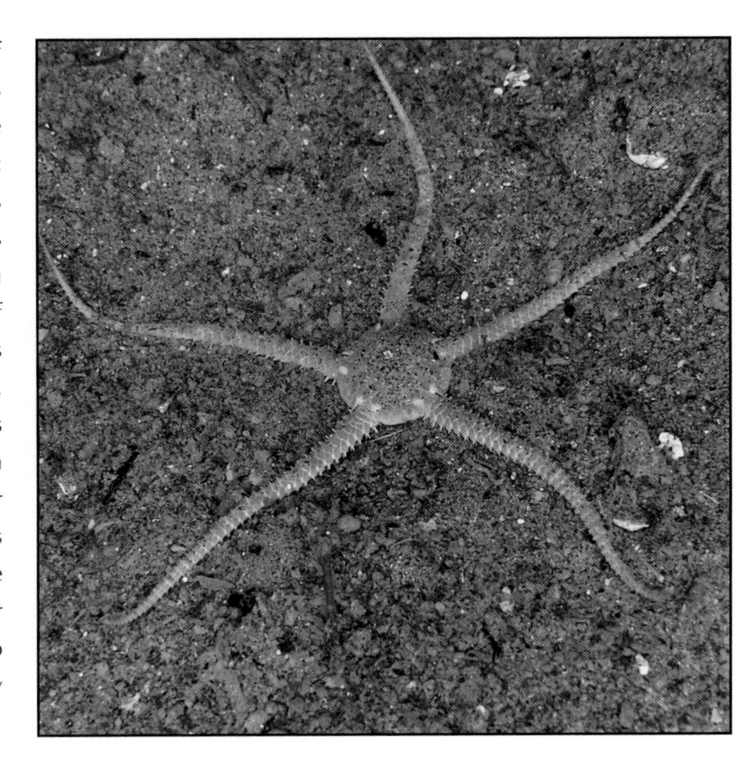

This small brittle star is one of the easiest to identify, thanks to the distinctive pair of white marks on the central disc near the base of each of its arms. It is found in silty areas and so is usually dusted with at least a fine covering of sediment, but these markings still tend to be clearly visible. The pattern of plates on its central disc, however, is often obscured. The overall colour is reddish brown while it is similar in form to its close relative the sand brittle star (opposite page). [Up to 14 cm across but usually smaller]

Sand burrowing brittle star - *Acrocnida brachiata*

As its name suggests, this brittle star lives almost completely buried within sandy seabeds. The whitish or greyish tips of its very long arms protrude from the surface of the sand in order to catch food drifting by. The wavy arms are covered in short spines and it is only these (and the tube feet if visible as in the photograph) that show you are looking at a brittle star and not some type of worm. Aggregations producing a sparse forest of arms are often found, and it is not clear which arms belong to the same individual. In any case, many brittle stars will only have one arm protruding while the others act as anchors. [Up to 40 cm across, but this measurement is very misleading. Usually, only a few cm of an arm or two are visible]

Single arm emerging from sand

Common (or edible) sea urchin - *Echinus esculentus*

Urchin with tube feet extended

Close-up of urchin, showing tube feet, spines and pedicellariae

Common (or edible) sea urchin - *Echinus esculentus*

Sea urchins grazing at the bottom of a gully lined with soft corals

The skeleton (also known as a test) of this impressive animal is very rounded and is usually bright red, though the numerous white spine attachment points tend to make it appear pink. Some individuals have a pretty purple hue. The abundant spines are strong, but quite short and thin relative to the overall body size in comparison to other urchins. The bottom photograph on the opposite page is an extreme close-up of an urchin's surface and it shows how the features are very similar to those of a starfish (see page 195), although with different proportions. The spines have the same basic form and the tube feet (top left and lower edge of picture) are almost identical, though of course longer in order to reach out beyond the spines when help with anchoring or movement is required. The pincer-like pedicellariae for removing "free-loaders" have long stems and are particularly noticeable. Without them, the slow-moving urchin would have even more trouble with accumulated debris and unwanted guests than would the relatively (!) athletic starfish. Urchins are powerful and omnivorous grazers, eating animals such as barnacles in addition to algae, and are capable of leaving virtually bare swathes across a rock face. They are abundant in many rocky areas (such as St Abbs, above) and can therefore form an important component of the ecology. Rock-dwelling urchins such as *Echinus* have a feeding mechanism known as Aristotle's lantern (it was he who first described it), which consists of a complex structure of plates and muscles that supports five chisel-like teeth used for scraping food from surfaces. This species is found all around Britain except for the eastern end of the Channel but, in some areas, it has suffered badly from the souvenir trade. [Up to 20 cm across]

Shore (or green) sea urchin - *Psammechinus miliaris*

Spawning

This urchin is much smaller than the common sea urchin but its spines are longer in relation to its body size, giving it a very spiky appearance (see bottom photograph for comparison). Overall colour is pale green and the spines have fetching purplish tips. As its name suggests, it lives on the shore and in very shallow water. It is often decorated and partly camouflaged by seaweed, shells and other debris trapped among the spines (see middle photograph). Like the common sea urchin, this species is an omnivorous grazer. Breeding is similar to that in many starfish, with the sexes separate and males and females releasing spawn from a good vantage point such as the top of a kelp frond (see top photograph). [Up to 6 cm across]

Decoration and camouflage, including the empty test of a smaller urchin

Common sea urchin on left; shore sea urchin on right

Common heart urchin (or sea potato) - *Echinocardium cordatum*

Heart urchin on surface of sand, small bivalve on right

Heart urchin burrowing, small bivalve on left

Test of dead animal without spines

Though very abundant, heart urchins are seldom seen because they burrow into a sandy seabed and spend most of their time 10 to 15 cm beneath its surface. The unusual individual in the top photograph was found crawling across the surface of the sand. The one in the middle photograph was in the process of burrowing back down again. Unlike the common sea urchin, the test (skeleton) is not round but oval, a shape better suited to the burrowing lifestyle. It also means there is a front and back end. Dense yellow/gold spines cover the test and all point backwards, giving the animal a furry appearance. The spines are used for burrowing, especially the ones shaped like spatulas on the under-surface. Heart urchins construct a channel that runs up to the surface of the sand where it forms a small but noticeable depression. Special, highly elongated tube feet stretch up the channel; these are used for respiration and for collecting the deposited particles on which the urchin feeds. Unlike urchins that are crunching grazers, heart urchins possess no Aristotle's lantern. A type of small bivalve mollusc is often found living in association with this species, as can be seen in the top two photographs. The tests of dead animals without the spines (see bottom photograph) are often found underwater or washed up on the shore; they are very fragile. [Up to 10 cm long]

Cotton-spinner - *Holothuria forskali*

This large sausage-shaped sea cucumber, found off south west and west Britain, is typically found crawling slowly over rocky seabeds. The one shown above was unusual in being on sandy pebbles, but its features came out more clearly than those of most cotton-spinners I photograph. Overall colour is usually black, but the yellow or pale brown on the underside may sometimes spread up over the rest of the body. Most of the body is covered with conical protuberances, except for the flatter underside which bears the tube feet used for locomotion. These features are reminders that you are looking at a relative of the starfish, rather than an exceptionally well fed and prickly slug! The cotton-spinner has short tentacles, rather than the long feathery ones of suspension-feeding sea cucumbers (opposite and page 226, top), and consumes silty deposits from the seabed. Having extracted the nutritious organic component, it leaves distinctive trails of undigested material that look like strings of large sandy beads. The name cotton-spinner comes from the animal's habit of producing long white threads from its rear end if molested (see smaller photograph). The threads are part of its internal organs and are extremely sticky, serving to confuse or entangle the attacker. [Up to 25 cm long]

Defences activated

Brown sea cucumber - *Aslia lefevrei*

The brown, leathery gherkin-shaped body of this animal is generally hidden in a rock crevice, but its large and highly branched tentacles extend out into open water. The tentacles are dark brown with some white edging often present. Mucus on the tentacles collects suspended matter, and each tentacle is in turn rolled up and inserted into the central mouth so that the food can be collected (see arm at "five o'clock" position), before it is then re-extended. It is very difficult to watch this process without visualising a child gradually sucking jam from their sticky fingers. This species can be abundant in some locations in the south west and north west of Britain. The cucumber in the photograph is surrounded by feather stars (page 227) and daisy anemones (page 53). [Tentacles up to 10 cm long]

Gravel sea cucumber - *Neopentadactyla mixta*

This sea cucumber has feathery white feeding tentacles which are extended above the gravel seabed in which it lives. The tentacles are attractively decorated with brown flecks which can be so dense that the overall colour becomes brown. There may also be striking brown and white patterns around the mouth. A small portion of the long, tapering body may be visible but the majority of its length is buried and hidden from view. If disturbed, the tentacles can be rapidly withdrawn, though individuals seem to vary greatly in their sensitivity. Like so many animals that rely on suspended food, they like areas with good water movement. [Tentacles up to 15 cm long]

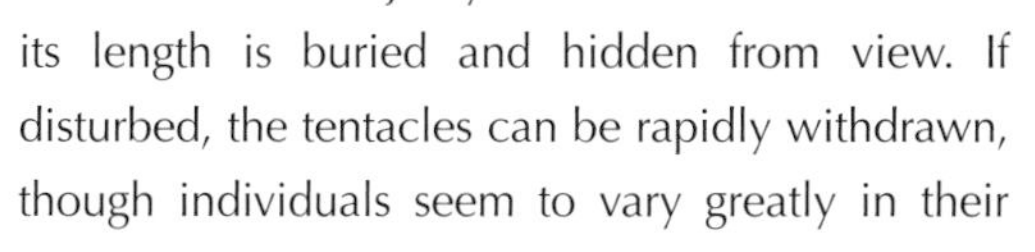

Pink spotted sea cucumber - *Psolus phantapus*

The bright orange patches on the creamy coloured tentacles of this sea cucumber species give it a very distinctive appearance. There are also orange or pink spots on the pale body. It is only found in the north of Britain and, even there, is not very widespread. This photograph was taken on a mud slope in Loch Duich where a large number of individuals were found dotted over the seabed. All but the tentacle crowns and the upper parts of their bodies were hidden in the mud. There are several similar sea cucumbers, not listed here, more widespread than *Psolus phantapus* but less distinctive and attractive. [Tentacles up to 15 cm long]

Labidoplax digitata

This creature is a sea cucumber even though it looks like a fat bald worm. It burrows in sandy and muddy seabeds and it is often only a section of the long speckly pink body that can be seen as it breaks surface. There are 12 small tentacles around the mouth at one end, usually hidden in the sediment, that are used to collect small particles of dead organic material. It is unusual amongst sea cucumbers (and echinoderms in general) in having no conventional tube feet. Found all around the west of Britain. [Up to 30 cm long]

Common feather star - *Antedon bifida*

This animal is common around most of the British coast (except the south east) and can occur in abundance. It has ten feathery arms of variable colour, though often red and white, while the overall pattern can produce a mottled or freckled appearance. On the underside of the small central disc there are twenty-five or so shorter claw-like appendages, called cirri, used for anchoring the feather star to the seabed, to kelp stems or to sponges (as in the photograph). Their grip is surprisingly tenacious, as divers who have found them clinging to equipment will know. Feather stars can swim by sweeping their arms up and down, or crawl slowly on the tips of the arms which are bent right over to hold the body away from the seabed. They spend the vast majority of their time, however, simply anchored in their chosen location where they engage in suspension feeding. The arms and their feathery branches are equipped with numerous tiny tube feet that catch floating food and flick it into grooves which run down each arm. This food is then transported by the beating action of tiny hairs down to the mouth in the centre of the body. Unlike starfish, urchins and brittle stars, feather stars have their mouth on the top side of their body. When not being used for either feeding or locomotion, the arms can be curled up over the body. Though often found singly or in small groups, feather stars are most noticeable when they form huge aggregations. These are most common in current-swept areas not exposed to heavy surf. A closely related feather star (*Antedon petasus*) is larger, neater looking, less freckly and has a greater number (50 or more) of the short claw-like cirri. [*Antedon bifida* up to 15 cm across; *Antedon petasus* up to 20 cm across]

Celtic feather star - *Leptometra celtica*

This species of feather star has a rather more elegant overall appearance than *Antedon bifida*. The best distinguishing feature, however, is the presence of very long white cirri (claw-like appendages) on the underside of its small central disc. These are used to cling to the seabed. The feathery arms are coloured in white, pale brown or red and are often banded. Usually regarded as a deep water animal, but is quite common in shallow water (around 15 metres) on the west coast of Scotland. [Arms up to 12 cm long]

Chapter 9
SEA SQUIRTS

Unlikely relatives

The sea squirts are a most deceptive group of animals. By living sedentary lives, and feeding by filtering plankton and other food particles from seawater, they appear similar to more primitive creatures such as the sponges. Amazingly, sea squirts are members of the phylum Chordata, the major animal group that contains such creatures as fish, birds and ourselves. The adult sea squirt gives little indication of this elevated status, but the tiny sea squirt larva resembles a tadpole and possesses a very simplified version of the backbone and nerve chord that are characteristic of its advanced relatives.

Larval secrets

Sea squirt eggs are released into the sea or brooded by the adult, depending on the species. When the larvae, or tadpoles, first hatch from eggs, they swim up towards the light. This is made possible by a light receptor (primitive eye) and a muscular tail which works well because of the stiffening provided by the rudimentary backbone (known as a notochord). Soon however, the larvae's behaviour changes, and they swim away from the light and down towards the seabed. Here, they attach themselves to a suitable substrate with adhesive pads on their front end, and completely lose their youthful mobility. The tail is reabsorbed, most of their more sophisticated features such as the notochord disappear, and an adult sea squirt results. Many species form colonies.

The adult form

The adult body is, in essence, a U-shaped tube surrounded by a tough and leathery tunic (this is why sea squirts are also known as tunicates). There is an efficient filtration and pumping system in the centre of the tube, based on the action of numerous beating cilia (tiny hair-like structures). The tube's ends are clearly visible as two openings or siphons: one for the intake of water and one for its expulsion. If a sea squirt is disturbed it can contract its body and water is squirted from the two siphons (hence the name). The body contains an intestine for processing the filtered material and a strange blood circulation system that is unique to sea squirts. Blood is pumped one way round the body for several seconds and the flow is then reversed for an equivalent length of time. Sea squirts are preyed upon by a variety of animals. Sea slugs, in particular, are often seen munching through the colonies. Some species protect themselves by having a highly acidic body wall which renders them unpalatable.

Yellow ringed sea squirt - *Ciona intestinalis*

Yellow ringed sea squirts with brittle stars and sea loch anemones

This widespread and abundant species is the archetypal sea squirt. It is tall, slim and cylindrical in shape but some individuals may be partially contracted and look quite squat. Its translucent tunic, through which the internal organs are usually visible, may be creamy white, pale yellowish green or pale orange-brown. The water intake siphon is at the top of the body with the outflow siphon just off to one side. Both siphons have a distinctive yellow rim and there are tiny red spots between each lobe of the rim that may be seen on close inspection. These sea squirts are fast growing and will rapidly take over freshly available habitats such as marina pilings and chains. They can form extremely dense aggregations, of up to 5000 squirts per square metre, but the individuals are not fused in any way so it is known as a solitary rather than colonial species. In natural and settled habitats, it is often found in small bunches or in ones and twos. When attempts are made to cultivate shellfish in disused dock basins, the yellow ringed sea squirt is a serious pest. As highly efficient filter feeders, each pumping up to 3 litres of water an hour and removing particles as small as a thousandth of a millimetre, the sea squirts tend to grab the lion's share of the food as well as the space. [Up to 15 cm tall]

Light-bulb sea squirt - *Clavelina lepadiformis*

These attractive small sea squirts live in bunches that may contain a few or dozens of individuals. The transparent tunic with its delicate yellow or white markings, and the visible internal organs, give the animal its name. Although a colonial species, individuals are free along most of their length and are joined to each other only at their base. The reddish eggs and developing larvae can sometimes be seen within the body cavity. The larvae are released in the late summer and the individual squirts then regress to leave "buds" which survive through the winter and develop into new squirts in the spring. [Up to 2 cm tall. Clumps up to 15 cm across are quite common]

Football sea squirt - *Diazona violacea*

An aptly named sea squirt that can form large ball-shaped colonies, although small groups such as the one shown here are also common. While there is superficial similarity to the light-bulb squirt (page 231), the football's tunic is less transparent and more milky. Individual members of football colonies are also more closely joined together. Small groups of light-bulb squirts in the top-left and top-right of the picture show how the appearances of the two species differ. The football sea squirt is usually restricted to deep clear water and is encountered much less often than the light-bulb squirt. I have mainly seen it off the west coast of Scotland. [Colonies up to 40 cm across]

Gas mantle sea squirt - *Corella parallelogramma*

This solitary species is quite small but is one of the most striking sea squirts, with an almost transparent tunic that is often marked by flecks of bright yellow or red. The intricate mesh-like structure that is responsible for pumping water through the body and filtering out food is clearly visible, as are other internal organs. The body tends to be rather flattened and usually has a squarish or oval outline, a little like a full hot water bottle. [Up to 5 cm tall]

Fluted sea squirt - *Ascidiella aspersa*

The most striking feature of *Ascidiella aspersa* tends to be that it looks dusty and in need of a good clean. This is because its grey tunic has a rough surface which tends to trap a layer of silt and other detritus. A more reliable identification feature is that the water intake siphon looks crimped and is at the top of the body, while the outflow siphon is positioned a third of the way down the side of the body. Although classified as a solitary sea squirt, it is usually found in clumps (see also page 266). These clumps may cover a large area of muddy seabed, with the squirts attached to pebbles or shells within the mud. [Up to 10 cm tall]

Red sea squirt - *Ascidia mentula*

Ascidia mentula is large and generally red or pink in colour, and this combination sets it apart from most sea squirt species. Its colour may be grey in the poor light of deeper water however. Rather than being attached to the seabed by its base like most sea squirts, it is joined on by a part of one side. The intake siphon is at the squirt's free end and is marked with eight white spots which are particularly obvious on deep red individuals. The outflow siphon is about halfway down the body and may be obscured by detritus and encrusting growth. Individuals are often found grouped together so what look like the two siphons of one squirt are in fact the intake siphons of two individuals whose outflow siphons are hidden. [Up to 15 cm long]

Phallusia mammillata

The largest British sea squirt and a truly solitary species, it is found in the south and west. Lives attached to stones, but often in muddy areas. It is rather plump in appearance and the milk-white or yellowish tunic is unusually thick for a sea squirt and appears quite stiff. Its body surface is covered with rounded, smooth swellings (hence *mammillata*). The water intake siphon is at the very top of the body, the outlet part-way down. Small anemones sometimes live on this sea squirt, taking the opportunity to get away from bottom silt and receive better access to passing food. [Up to 15 cm tall]

Club sea squirt - *Aplidium punctum*

This sea squirt forms close-knit colonies where individuals are barely distinguishable from each other. Each colony appears to have a bulbous head on a thinner stalk, and contains approximately forty or more individuals. Close examination reveals that each individual is marked with a tiny dark orange spot that stands out against the generally whitish head. Several colonies are often bunched close together (as shown here), attached to rocks, stones or seaweed. The stalks of the colonies are fairly pale. Colonies of a very similar species, *Morchellium argus*, have more reddish, shorter stalks. Each individual in the colonies has four red spots instead of the one in *Aplidium* but this is hard to make out underwater. [Colonies, including stalk, up to 4 cm long]

Sidnyum elegans

This species forms distinctive cushion-like colonies with an overall deep pink colour. Numerous water intake siphons cover the colony and each has an attractive white frill around it. The larger (and fewer in number) outflow siphons arise from canals which appear to be draped across the colony's surface. Found mainly in the English Channel, it can be common in some areas. The photograph on the right shows a group of *Sidnyum elegans* colonies on a leg of the bridge over the Fleet in Dorset. There are two of the large solitary sea squirts *Phallusia mammillata* (opposite) near the bottom of the leg. [Colonies usually up to 6 cm across, but sometimes larger]

Leathery sea squirt - *Styela clava*

The stalk of this tall sea squirt gives it a distinctive club-like shape, although the stalk itself can be hidden by creatures living nearby. Its blotchy brown surface has a slightly "padded quilt-like" appearance and the two siphons, which can bear broad stripes, are right at its top end. It is found singly or in small groups; the photograph shows a row of squirts surrounded by feather stars (page 227). *Styela clava* is a native of the Pacific Ocean and is thought to have been brought to our shores on the hulls or in the ballast tanks of ships. It is common in many sheltered areas on the south coast and is also found further north. I have seen it in marinas in East Anglia. [Up to 12 cm tall]

Gooseberry sea squirt - *Dendrodoa grossularia*

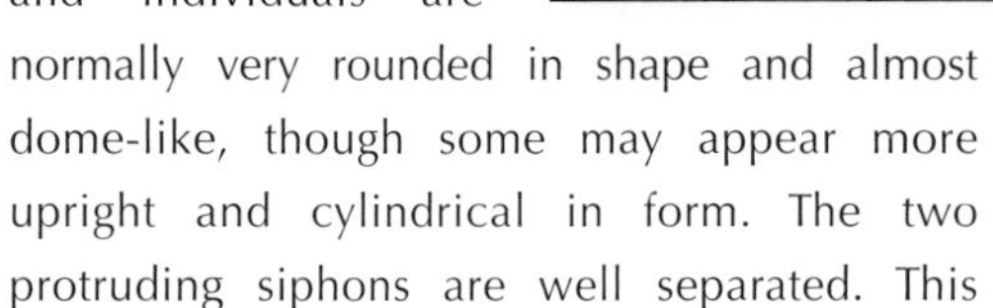

This species is defined as solitary because it can live alone, but it is most likely to be seen in aggregations on the walls of rock gullies, under overhangs or on stones. Such groups result from the settling larvae's gregarious behaviour, rather than from budding. The squirts' usual colour is a cherry or orangey red and individuals are normally very rounded in shape and almost dome-like, though some may appear more upright and cylindrical in form. The two protruding siphons are well separated. This species is readily consumed by predatory sea slugs which can sometimes be found chewing their way hungrily through the aggregations. [Up to 2 cm tall]

Orange sea squirt - *Stolonica socialis*

These sea squirts are known as colonial because the individual squirts arise from a joint base structure a little like "runners" in plants (connections like these are called stolons, hence the scientific name). The aggregations, however, do not appear any more closely-knit than those of the gooseberry sea squirt (above) which are not linked in this way. Orange sea squirts are a little taller in shape and less rounded than gooseberry sea squirts as well as being a different colour. Orange sea squirts are also usually seen in deeper water, while gooseberry sea squirts can be found in the shallows. [Up to 2 cm tall]

Star sea squirt - *Botryllus schlosseri*

Close-up of colony

Colonies of star sea squirts are so closely organised that they bear hardly any resemblance to the standard sea squirt form. A colony resides in a common gelatinous tunic, while groups of three to twelve individuals within the colony each produce one of the characteristic star-shaped patterns that cover its surface. Within these groups, the individuals have separate intake siphons but share a common outflow opening in the centre of the star. The basic tunic is usually a dark colour while the stars are a contrasting yellow. Colonies may appear flat or bulbous in their overall shape and are found encrusting rocks and seaweed. Reproduction can be sexual or achieved by budding. [Colonies around 10 cm across are common]

Chapter 10
FISH

One of the first things that strikes someone snorkelling or diving in British waters for the first time is the large number and variety of fish to be seen. While the classic fish shape is represented by species such as bass, grey mullet and pollack, many of the other fish seen in shallow water have different shapes which reflect their bottom-dwelling existence. Adaptations to this lifestyle can include a flattened and enlarged head, with upward-facing mouth, like sea scorpions, weevers and angler fish. Such fish spend a great deal of time lying stationary on the seabed, relying on camouflage to avoid predators and enable them to ambush their prey. Other species have become highly elongated so they can hide amongst seaweed (pipefish) or slip into the narrowest of crevices (conger eels). Still others have thin plate-like bodies, staying upright to sneak up on their prey (John Dory) or lying on their sides on the seabed (plaice, sole).

Fins: their layout and function

Despite the variety in fish body shape, it is quite easy to see that all the various species conform to the same basic plan. A good clue is the fins which, despite unusual appearances or adaptations to different functions, are generally laid out in the same pattern (see below). The tail fin provides propulsion and acts as a rudder, while the dorsal and anal fins help with stability by preventing body roll and also act as pivots when the fish is turning. They are known as the

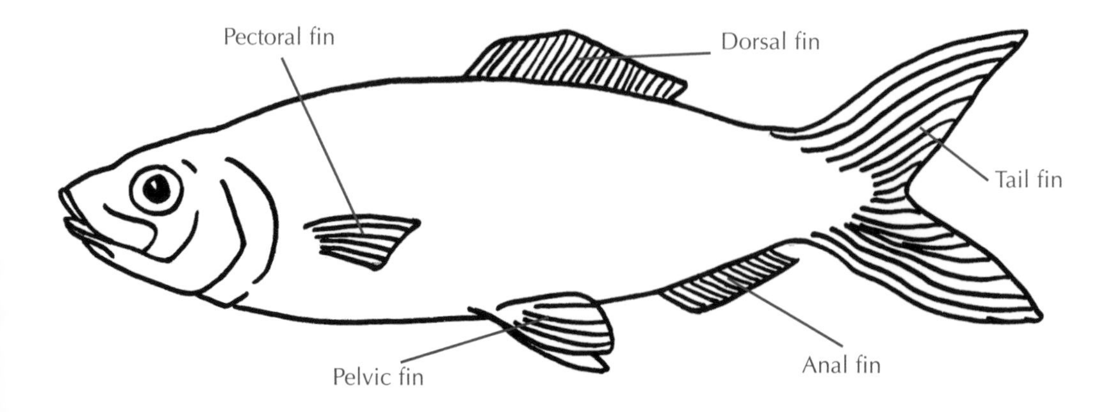

unpaired fins though there may be one, two or even three dorsal fins. The paired fins, pectoral and pelvic, correspond (as their names imply) to the front and hind limbs of land-living vertebrate animals such as ourselves. Pectoral fins are used to help with all sorts of manoeuvres such as turning, braking, rising and falling. The pelvic fins vary considerably in their position from one species to another; they tend to help the pectorals with manoeuvering and maintaining stability. In addition to these basic functions, fins may be used for catching food, repelling predators, mating, courtship display and assisting with camouflage.

Varied approaches to breeding

Many open sea fish take virtually no care of their young and adopt a "squirt and hope" strategy. A female and male shed vast numbers of eggs and sperm into the water near each other, and seem to make little effort to ensure that fertilisation occurs. Parental attention thereafter is non-existent, and the developing embryos and fish larvae are left entirely to their own devices amongst the plankton in open water. It's not that the parents "don't care", it's simply that they put their energy and resources into quantity rather than quality. Such resources can be substantial, as demonstrated by the weakened state of many fish after breeding. In some species that use this strategy, adults die immediately after their one and only spawning. Many of the shallow water fish familiar to divers use an alternative approach to "squirt and hope". They produce far fewer eggs and make much more strenuous efforts to ensure fertilisation and the survival of the resulting young. It seems that a strategy suited to the open sea is less successful in coastal waters where predators are more numerous and physical conditions are harsher. Abandoned eggs and larvae could easily be washed into less suitable habitats such as estuaries or deeper water. Interestingly, it is very often the male of common coastal species that takes on the main burden of parenthood. Their nest-building and egg-guarding activities make an interesting spectacle for the observant diver.

Skeletons of bone or cartilage

The major division in the fish world is that between those with skeletons made entirely from cartilage (sharks and rays) and those with bony skeletons (all the rest). The cartilaginous fish have other characteristic features, such as exposed gill slits and skin covered with tooth-like scales. Their fins are much thicker and more fleshy than those of bony fish, and may also be used slightly differently. The pectoral fins are usually large and, because a buoyancy organ is lacking, act as hydroplanes to provide lift for swimming. Rays have taken this to the extreme and have huge pectoral fins used as wings. Despite these differences, however, the sharks and rays share the same basic fin layout already described. The bony fish have complex and hardened skeletons, similar to those of other vertebrates like ourselves, and their thinner fins are supported by fin rays. They are far more numerous than the cartilaginous fish, with over twenty thousand species versus seven hundred or so. An important characteristic of bony fish is their more sophisticated buoyancy control system, that uses an adjustable gas-filled organ called a swimbladder. By adding or removing gas from the swimbladder, in a similar way to a diver using a buoyancy compensator, neutral buoyancy can be achieved so that energy is concentrated solely on forward motion. However, many of the bony fish species that live on the seabed, such as blennies and flatfish for example, do not actually need this facility and have lost their swimbladders.

Basking shark - *Cetorhinus maximus*

Basking shark feeding at surface

This magnificent and rather mysterious animal is regularly seen around our coasts, although sightings can sometimes be frustratingly elusive. Basking sharks are found virtually worldwide in cooler seas and, though they can crop up all around Britain, most sightings are on western coasts. Particular "hot-spots" seem to be Cornwall, the Isle of Man, and the west of Scotland. Basking sharks can reach up to 12 metres long (the equivalent of two large estate cars end-to-end!) and weigh 7 tonnes. They are the **second largest fish in the world**, only the whale shark of tropical seas is bigger. The body is dark grey or greyish brown and has the overall shape of a typical shark, on a large scale! The unusual long snout, which appears even longer in young sharks, is most striking when the enormous mouth is closed. Five very long gill slits run almost completely around the body. Basking sharks have negligible teeth and, when swimming with their huge mouths gaping wide open, **filter plankton** from the seawater with specially adapted gill rakers. They are far from indiscriminate filter feeders, however, and they actively track and choose to feed in patches where there is a rich density of zooplankton. They can be seen feeding at the surface in the late spring and summer, particularly in calm sunny

Basking shark - *Cetorhinus maximus*

A snorkeller with video camera gives an idea of scale

weather. This habit explains the name **"basking"** and, while swimming here, the dorsal fin, tail and even the very tip of the snout may all break the surface. It is then possible to carefully enter the water in snorkelling gear and watch these spectacular creatures in their own habitat. Whether in the water or on a boat, it is crucial not to harass the sharks and to give them space, following the **Basking Shark Code of Conduct**. As a **protected species** in British waters, it is illegal to injure, kill or recklessly disturb them. Basking sharks give birth to **live young**, just a few at a time. Youngsters are about 1.5 metres at birth and are rarely seen before reaching at least 3 metres. Research using satellite-linked tags has shown that basking sharks do not hibernate as was once thought. There is still much uncertainty about their biology (the length of pregnancy for instance) but what is certain, is that their late age of maturation (possibly up to 20 years for females) and small number of young make them extremely **vulnerable** to overexploitation. They have been fished for their oily liver, which provides the shark with buoyancy in the absence of a swim bladder, and for their fins which (tragically) are extremely valuable as a base for soup. The species is officially listed as "vulnerable" and has been protected internationally under Appendix II of CITES (the Convention on International Trade in Endangered Species of Wild Fauna and Flora) since 2002. [Up to 12 m long, most sharks seen around Britain 3 to 9 m]

Smallspotted catshark (or lesser spotted dogfish) - *Scyliorhinus canicula*

This catshark (often known misleadingly as a dogfish) is much the most common shark encountered by divers around Britain. It has the clear shark characteristics of a mouth positioned on the underside of the head, and obvious gill slits just in front of the hydroplane-like pectoral fins. The dorsal fins are small, rounded and positioned well back on the body, however, and thus bear little resemblance to those of the larger sharks. The body is a greyish brown in colour with numerous dark spots, and a white belly. Tooth-like scales that cover the catshark have spines that point backwards. The skin, once used as sandpaper, feels quite smooth if stroked from nose to tail and very rough in the opposite direction. Catsharks hunt at night, eating a wide variety of prey that can include crabs, whelks and bottom-living fish such as gobies, dabs and gurnards. They apparently rely heavily on their sense of smell when hunting and may slavishly follow a scent trail, even when their prey has turned and swum right back past them. They can also detect faint electrical fields produced by the muscles of hidden prey. Groups of female smallspotted catsharks can often be found resting in caves or gullies (opposite page, top photograph). The main reason for them doing this is apparently to take refuge from aggressive courting males. After mating, females lay their distinctive pale brown egg capsules, known as "mermaids' purses". These have long tendrils on the corners and, as they are laid, the mother fish will swim round and round a clump of seaweed, a sea fan or similar anchorage, so that they become well attached (opposite page, bottom photograph). Each capsule contains a single embryo and a miniature catshark emerges after nine months or so. [Fish up to 80 cm long, egg cases are about 7 cm long]

Smallspotted catshark (or lesser spotted dogfish) - *Scyliorhinus canicula*

Four females in a refuge

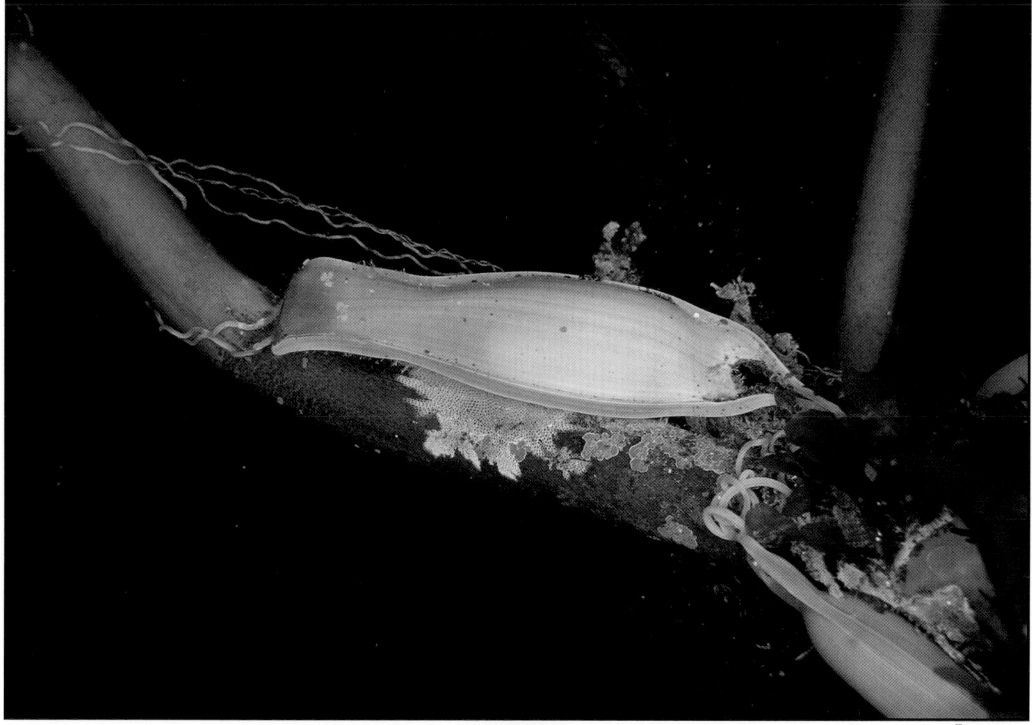

Egg cases

Thornback ray - *Raja clavata*

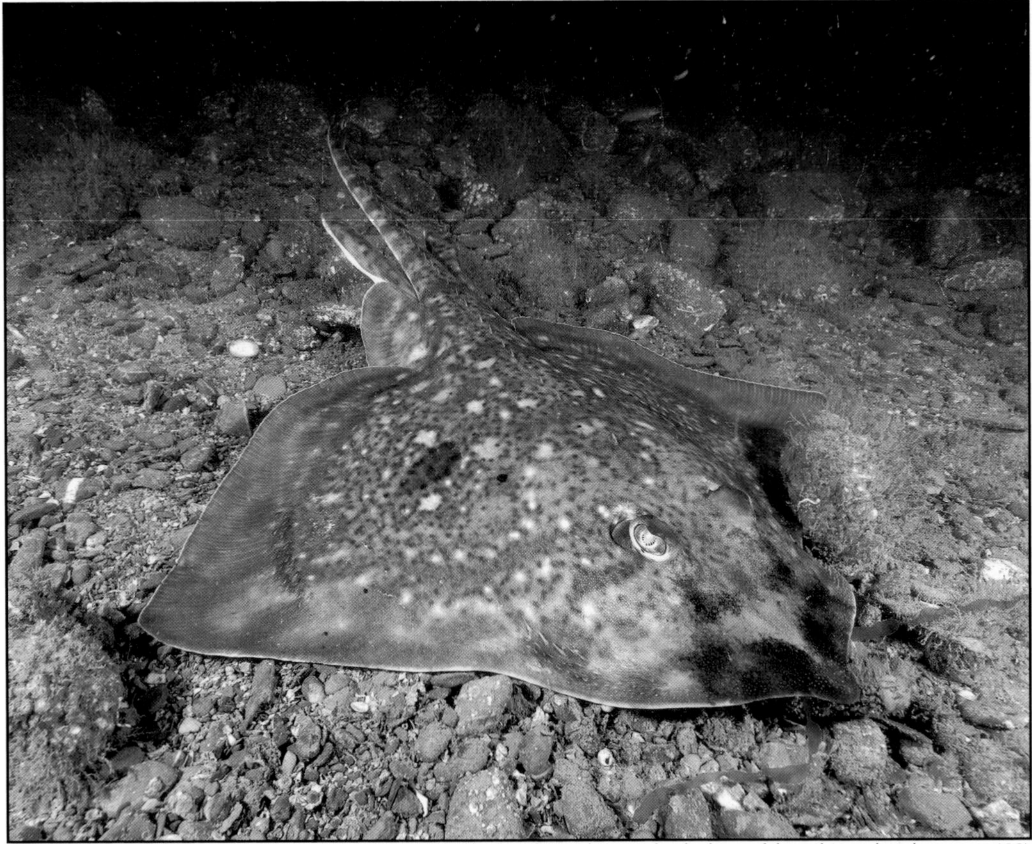

Male ray; the long claspers used for transferring sperm to the female can be seen by the base of the tail (see also photo on p.139)

The thornback (or roker) is the most commonly encountered ray around Britain, being found in relatively deep water and also in very shallow bays and estuaries. It is usually a blotchy brown or grey and, as the name suggests, bears an array of large thorn-like projections on its back and tail. Some "thorns" have an obvious plate-like base. Rays and skates belong to the same group of fish as sharks and, similarly, have a skeleton made of cartilage rather than hard bone. With their flat diamond-shaped bodies, rays can be thought of as "squashed sharks". The greatly enlarged pectoral fins form the "wings" which give them a wonderfully graceful flying motion when swimming (top photograph, opposite). Like most rays, the thornback spends much of its time lying motionless on the seabed where its body shape, along with the habit of covering itself in sand, may make it very difficult to spot. Fish usually take in water for breathing through their mouths, but rays can use the breathing holes (spiracles) behind the eyes on top of their head; they thus avoid taking in too much sand. Like the catshark, the thornback ray lays its eggs as "mermaid's purses" but these are black rather than brown and lack tendrils (bottom photograph, opposite). It feeds on a variety of bottom-living animals, especially crabs and shrimps. The related "common" skate (now regarded as a complex of two *Dipturus* species) has been fished to virtual extinction around most of Britain. Thornback ray populations now seem to be declining too. [Thornback up to 1 m long, skate can be well over 2 m long]

Thornback ray - *Raja clavata*

Female ray swimming

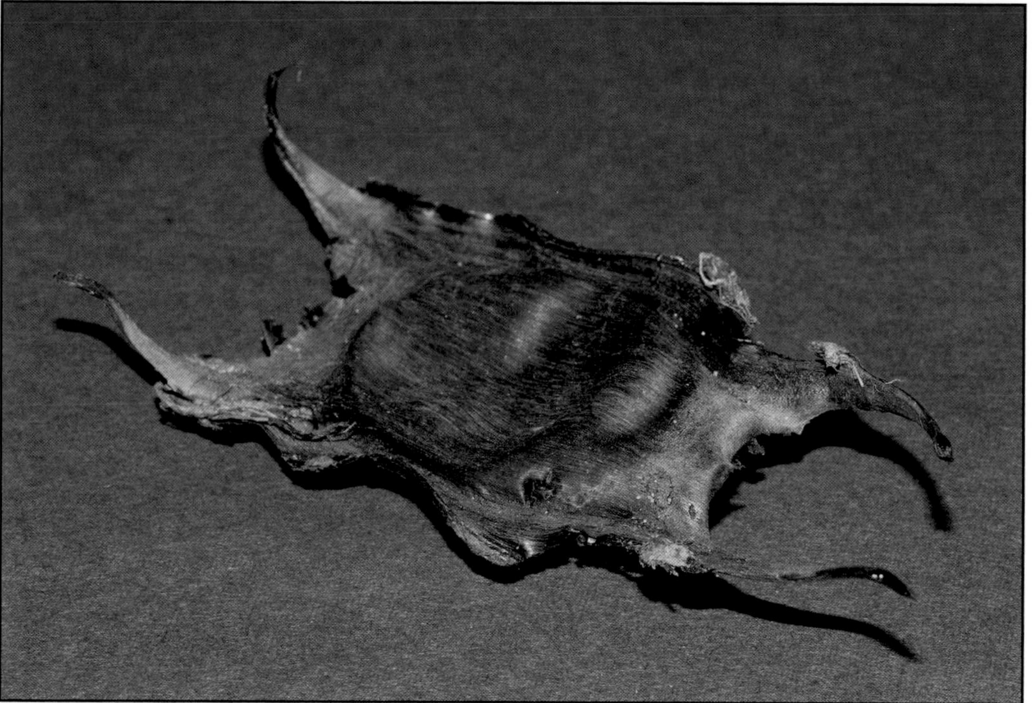

Washed up ray egg case

Conger eel - *Conger conger*

The scientific name is uniquely memorable and its owner is pretty unforgettable too. It is a highly elongated fish with a powerful snake-like body, grey-brown to grey-blue in colour with a paler underside. There are no scales and its skin is smooth. The dorsal, tail and anal fins are merged to form a single fringe that runs from just behind the head, right round the pointed tail, to underneath the belly. Congers hide in rocky holes and crevices however, so it is usually only the head that is seen, with its large mouth and distinctive snout bearing a pair of tubular nostrils. Shipwrecks, with all their nooks and crannies, are popular residences. Congers are formidable predators and will slip out of their lair at night to hunt fish, including smaller congers, and large crustaceans such as crabs and lobsters. The latter

Conger eel - *Conger conger*

Prawns with conger

are seized and may be battered against rocks before being swallowed. Surprisingly, this doesn't seem to stop lobsters and prawns (see photograph right) often sharing the crevices that congers use as home. Congers do not breed in our coastal waters but travel vast distances to spawn in the deep mid-Atlantic. Their bodies change as they approach the breeding grounds with teeth falling out, gut degenerating and the gonads becoming so greatly enlarged that they make up a third of body weight. They die after spawning and their larvae drift back into coastal waters and then turn into young eels. Congers have incredibly strong jaws and a reputation for ferocity but remain docile underwater unless provoked, so anglers have more to fear than divers. [Up to 2 m long and occasionally even larger]

European (or common) eel - *Anguilla anguilla*

A slender fish with an amazing life story, it looks similar to the conger but is usually much smaller. As in the conger, the dorsal fin merges with the tail and anal fins to form a single fringe. This starts well behind the pectoral fins, while in the conger it starts very close to the pectorals. The face also appears different because this eel's lower jaw is more prominent than the upper, while a conger's jaws are of equal length or the upper is slightly longer. The European eel spends most of its life in rivers and estuaries feeding and growing. Here, it has a green or brown back with a pale yellowish belly and is known as a yellow eel. As it becomes mature, the sides darken to almost black, the belly turns silver and it is described as a silver eel. Silver eels stop eating and go on a long journey down their rivers, through coastal waters and out into the deep Atlantic (mainly the Sargasso Sea) where they spawn. The adults die and the eggs hatch into planktonic leaf-shaped larvae which drift back on the Gulf Stream into European waters and develop into young eels (called elvers). These swim up rivers where they then become yellow eels and the full cycle is complete. [Males up to 50 cm long, females up to 1 m. Most eels seen are much smaller]

Sprat - *Sprattus sprattus*

Shoal of young sprat with a few young herring

The sprat belongs to the silvery shoaling herring family. Other members include the herring itself, the pilchard (called sardine when young) and the anchovy. Common features include a slender body, a single short dorsal fin and a forked tail. Their shape is perfect for rapid swimming in open water and their colouration provides optimum camouflage as it makes them hard to pick out when viewed from any angle. The different species are extremely difficult to identify underwater, as all that is seen is a rapidly-moving mass of silver. A shoal of these fish swirling near the surface is one of the most fabulous underwater sights. The sparkling, swooping maelstrom is usually accompanied by predatory fish darting up from below and sea birds diving in from above to take their share. After a shoal has been subject to such an attack, the water is filled with detached scales looking like tiny scraps of silver paper. The shoal in these pictures consisted mainly of sprat, but there were a few young herring too; identification being based on close examination of fin positions on fish from the whole roll of film. Such mixed shoals of fish are known as "whitebait". The photographs were taken in only 1 metre of water, and within 5 metres of the shore, on the North Sea coast of Scotland. The shoal had been trapped in a rocky inlet by a gang of young saithe (see opposite bottom). Sprat and all members of the herring family feed on small planktonic animals which they sieve from the water as they swim along. In turn, they are an important food source for many large fish, seals and man. The European herring fishery has been of huge historic importance with many coastal towns growing up where the fish could be easily caught and landed. [Sprat up to 15 cm and herring up to 40 cm long, but it is much smaller fish that are usually seen near the coast]

Sprat - *Sprattus sprattus*

Shoal of young sprat, with a few young herring

A young saithe preying on the shoal

Angler fish - *Lophius piscatorius*

This very unusual fish, gastronomically known as a monkfish, is superbly adapted to its habitat and lifestyle so, considering its potential size, it can be surprisingly difficult to spot. The whole body is flattened, particularly the enormous broad head, and the tail appears small by comparison. Overall body colour is usually brown or greenish-brown with a white underside. Lying on the bottom, its mottled pattern gives good camouflage and the outline is further broken up by the fringe of small flaps of skin round the base of the head. The angler's name, of course, stems from its cunning means of attracting food. The foremost spiny ray of its dorsal fin is separate from the rest of the fin and serves as a "fishing rod". This "rod" can be moved around and the fleshy tip acts as a lure, inviting inquisitive creatures perilously close to the huge upturned mouth. The prey is then engulfed, a process aided by the inrush of water as the angler opens its jaws. Large inward-curving teeth make escape impossible. All sorts of bottom-living fish (including flatfish, gurnards, rays and conger eels) and other animals are eaten, even diving birds and gulls. The angler fish spawns in very deep water, laying its eggs in sheets which can form huge rafts up to 10 metres long and have occasionally been mistaken for sea monsters. A sheet typically contains around a million eggs. Hatchlings live freely in open water until they are around 8 cm long, and then take up life on the seabed. Extra-long fins help keep the youngsters afloat, but these have receded by the time the fish come to settle. The angler shown here was found on a ship wreck off Sunderland, the orange in the foreground is one of its rusty plates. [Up to 2 m long but it is much smaller individuals that are usually seen]

Cod - *Gadus morhua*

Young cod

Adult cod in the National Marine Aquarium at Plymouth

The typical cod family characteristics are a sensory chin barbel used in the search for food and a distinctive outline created by three closely positioned dorsal fins and two anal fins. Several closely related fish share some or all of these features. Although it is the best known member of this family, the cod itself is generally encountered much less frequently underwater than other family members such as pollack, saithe, bib and ling. It is very young cod that are most often seen (main photograph). They have an attractive chequered pattern of speckling and can be found darting amongst seaweed in shallow water. Adult cod are heavily built and are also distinguished by the very pale, curved lateral line which is surprisingly obvious underwater (see smaller photograph). The background flank colour of adults can be mottled greenish brown through to dirty yellow or even red. Adult cod are very difficult to approach and will dive into a crevice, or swim off rapidly, at the slightest provocation. This is why I have had to rely on an aquarium photograph. Female cod are renowned for producing several million eggs at a time and theoretical calculations have shown how quickly the Atlantic Ocean would become solid cod if all the eggs were to survive! They don't of course and, ironically, populations of cod have now been so seriously depleted by overfishing in most parts of its range that it is classified as a vulnerable species. Opposite to many fish, they move into deeper water in the summer, which further contributes to them being seen infrequently. [Up to 1.5 m long]

Pollack - *Pollachius pollachius*

Adult

Seen either singly or in (usually small) groups, the pollack is a very common fish in British waters. In many areas, it will be seen on every dive over shallow rocky ground in the summer months. The pollack is a member of the cod family and conforms to their typical pattern of three dorsal and two anal fins, though there is no chin barbel. Adults (main photograph, this page) are generally dark green along the back and more silvery on the sides and belly. The lateral line is usually obvious and takes a marked upward curve by the pectoral fins, in contrast to the straighter lateral line of the very similar saithe (page 254). Juveniles, which can often be seen very close to the shore, are more darkly coloured in green or brown and some may even be a fetching crimson and gold (smaller photograph, this page). With only a little patience, adult pollack can be watched feeding on lesser sand eels (page 270). Hunting individually or in small groups, they will lurk close to the seabed, gazing watchfully at the sand eel shoal swirling above them (top photograph, opposite page). Suddenly darting up through the shoal, they will grab any unwary fish before returning to near the bottom and waiting for the shoal to re-group. Juveniles can be seen using an identical technique on groups of mysid shrimps or two-spot gobies (page 300). The bottom photograph on the opposite page shows a pollack with a lamprey attached. Lampreys are primitive jawless fish that feed by attaching themselves to living fish with their mouths, rasping through the host's skin and then sucking their blood. The damage done to the pollack's skin is clear to see. [Pollack are usually up to around 50 cm long, larger fish sometimes seen near wrecks. Juvenile shown above is about 8 cm]

Juvenile

Pollack - *Pollachius pollachius*

Pair of pollack hunting sand eels

Pollack with lamprey attached. The lamprey is about 25 cm long

Saithe (coley or coalfish) - *Pollachius virens*

The saithe is a close relative of the pollack and the two can be very hard to distinguish, particularly if they are flashing by in hot pursuit of their prey. The saithe, which is relatively more common in the north of Britain, has jaws of approximately equal length in contrast to the jutting lower jaw of the pollack. The other most noticeable difference is the lateral line which, in the saithe, is approximately straight with a much more gentle curve over the pectoral fin than in the pollack. Unlike the loosely packed pollack groups, saithe are often seen in large and more tightly packed shoals. The photograph here shows a dense shoal of quite large saithe in Loch Carron; also see photograph of a young saithe preying on sprats on page 249 (bottom). [Up to 1.3 m long, but it is much smaller fish that are usually seen]

Five-bearded rockling - *Ciliata mustela*

Aquarium photograph

If a dark, slender (but good-sized) fish wriggles away when disturbed on the shore, it will often be one of these rocklings. The five-bearded rockling's name derives from its impressive set of face barbels which are thought to be used in locating its prey of small fish and invertebrates such as crabs. There is one barbel on the chin, a pair just above the top lip and a further pair above these in line with the eyes. Body colour is dark brown with a slightly paler underside. There are other species of rockling in our waters including the shore rockling (*Gaidropsarus mediterraneus*) which can be common on rocky shores in the south west; it has three face barbels. Both the five-bearded and shore rocklings can also be found below the shore, along with other rockling species. Any rocklings I have seen while diving have always been too far back in their crevices to photograph or identify clearly. All rocklings have an intriguing dorsal fin arrangement. They have two of them and, while the rear fin is a normal long and straight edged one, the front fin consists of a single long ray and then a fringe of very short rays which lie in a groove and can be very hard to see. If watched closely, this little fringe can be seen vibrating; opinion seems to be divided on whether this system helps to detect food. [Five-bearded rockling usually up to 25 cm long; this and other rocklings can be larger]

Bib - *Trisopterus luscus*

The bib is a common member of the cod family and has the typical three dorsal fins. Its body has quite a deep shape and is coppery coloured with an attractive banding that is usually, but not always, visible. There is a dark spot at the base of each pectoral fin. These features, in combination with a single barbel underneath the chin, should make the bib very easy to distinguish from all fish other than the poor cod (below). Bib are frequently seen inside and near wrecks in small or large groups. Shoals of very small bib, often found over the sand near wrecks, will approach and even surround divers whilst in search of any small pieces of food that they stir up from the bottom. Bib feed on crustaceans, such as shrimps, and molluscs with older fish eating other fish and squid too. [Up to 50 cm long but usually no more than 30 cm]

Poor cod - *Trisopterus minutus*

Belonging to the same genus as the bib, the poor cod is a close relative and possesses a very similar appearance. It shares features such as the protruding upper jaw and the long barbel on the lower. The poor cod can be distinguished from the bib by its shallower body shape, lack of banding and generally smaller size. It may be seen individually or in quite large groups. Shoals of small fish seen near wrecks are often a mixture of both bib and poor cod. [Up to 25 cm long]

Ling - *Molva molva*

This long, slim fish is a distinctive member of the cod family. It has the sensory chin barbel worn by most family members but lacks the typical three dorsal fins. Instead, the ling has one short dorsal fin, with a second much longer one behind. The overall colour is generally a marbled greenish brown; and the dorsal fins have an attractive pale edging which is most obvious when the fish is swimming (see smaller photograph). Youngsters are sometimes decorated with pale purplish iridescent lines. Ling are usually found peering out from rocky holes and crevices; or resting on the seabed just outside their hiding place (main photograph).

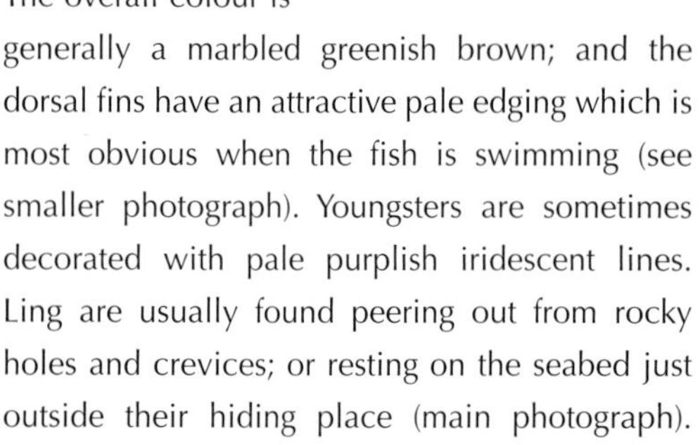

They will often sit facing towards it and, if disturbed, will dart back in and watch developments from the safety of their lair. Ling are most commonly seen around Scotland and the far south west. They can be a frequent sighting on wrecks in deep water. [Up to 2 m long. The largest fish normally reside in very deep water below 100 m]

Sand smelt - *Atherina presbyter*

Sand smelts are generally found in shoals which are much more loosely grouped than the tightly packed shoals of fish such as sprat (page 248), herring and mackerel. They are slim with two dorsal fins positioned quite far apart. The back is a very light silvery green and the lower half of the body is even paler. A shiny line runs down the middle of each flank. Their usual habitat is over sandy seabeds but sand smelt also seem to have a particular liking for piers; I have seen large shoals beneath the piers at Swanage in Dorset and Trefor in North Wales. The shoals at Swanage can often be watched being harassed by large bass (see page 271). [Up to 20 cm long]

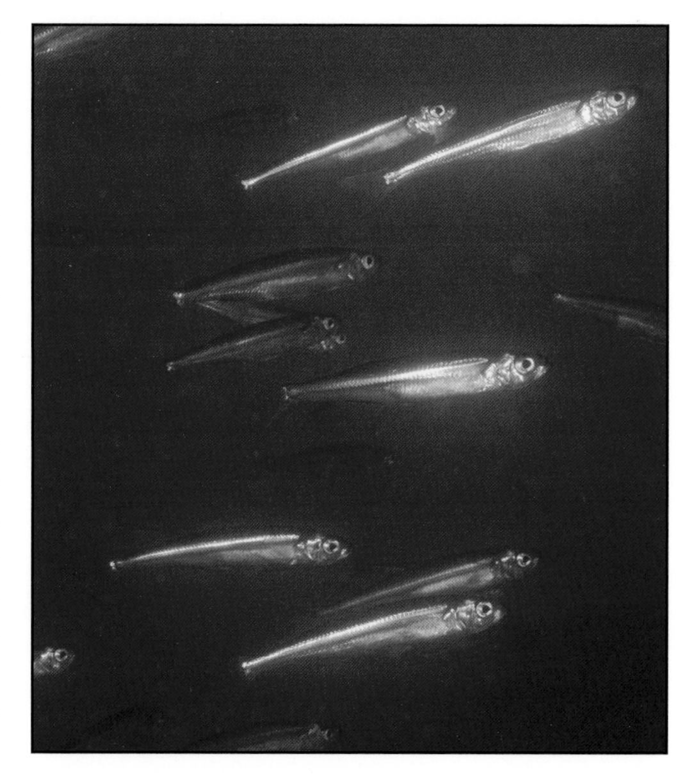

Fifteen-spined stickleback - *Spinachia spinachia*

This distinctively slender fish has an elongated snout and a very thin tail stem. A single dorsal fin is set well back towards the rear, behind the 15 small spines which give the fish its name. The male builds a nest, up amongst seaweed fronds, where he then guards the eggs laid by the female. Fifteen-spined sticklebacks are often found amongst eel-grass. They are usually seen singly, but the pair in this photograph were very keen to keep together as they hid amongst loose clumps of seaweed floating over the sandy seabed. This species is also known as the sea stickleback because, unlike other sticklebacks, it is only found in the sea. [Up to 20 cm long]

John Dory - *Zeus faber*

John Dory (with small parasitic crustaceans visible on its flank at the rear)

A highly distinctive fish, the John Dory has a body shaped like a flat oval plate on edge. Other notable features are the large head with its mournful expression, the very long dorsal fin rays and a single dark spot in the centre of each flank. The thin body virtually disappears when viewed head-on or tail-on (see photograph opposite), a characteristic used to good effect when approaching unsuspecting prey or avoiding predators. While not a rapid mover, the John Dory shows great manoeuvrability when stalking its prey, swimming while tilted at all sorts of angles and even upside down. At the same time, stripes on its flanks can alternately recede and intensify which, when coupled with the thin outline, makes its approach even harder to spot. Its jaws have a special bone construction which makes them highly protrusible. Once close to its unsuspecting prey, the mouth shoots out to engulf the victim. Prey is usually small fish such as sand eels, herring and pilchards. In the shallow water where encounters with divers tend to occur, they are often seen hunting the two-spot gobies (page 300) that hover in patches of seaweed. Legend has it that the dark blotches on each flank are St Peter's fingerprints, left when he took a coin from the fish's mouth to pay his tax. [Up to 60 cm long but usually much smaller]

John Dory - *Zeus faber*

Turned head-on, the slender outline of the John Dory is much harder to see

Greater pipefish - *Syngnathus acus*

Pipefish look like straightened and elongated versions of sea horses, to which they are closely related. Pipefish are like sea horses in having no scales. They have a series of jointed, bone-like rings which encircle the body from just behind the head down to the tail, giving most species a rather rigid, armoured appearance. The small mouth is positioned at the end of a slender snout and is used to suck in the tiny animals on which pipefish feed. Pipefish may move around by using a snake-like body action or, like sea horses, by rapid fluttering movements of their fins. The characteristic features of *Syngnathus acus* are its rough body with angular cross section, the very long snout taking up more than half the total length of the head and the bump on top of the head behind the eyes. It is usually found amongst seaweed or eel-grass where its shape and colour provide effective camouflage. The fish species that live in shallow water generally give their eggs good care, but pipefish take this to the extreme. The female lays her eggs into a special brood pouch on the belly of the male, where the young develop until emerging as miniature adults. [Up to 50 cm long]

Deep-snouted (or broad-snouted) pipefish - *Syngnathus typhle*

The name says it all. This species is very similar in form to its close relative the greater pipefish, but it has a much deeper snout. This means that there is no dip (like a forehead) in front of the eyes and the head is almost the same depth throughout its length. Colouration is an overall pale green or brown, darker above and lighter below, with some subtle patterning. Sometimes known as the "seagrass pipefish", it is often found in eel-grass where its shape, colour and habit of swimming and hovering vertically can make it very difficult to spot. [Up to 35 cm long]

Worm pipefish - *Nerophis lumbriciformis*

The small worm pipefish is rarely seen while snorkelling or diving but can be common in pools and among seaweed on the lower shore. The common name is appropriate because its body has a round cross section and is very slim and smooth. There is a small dorsal fin but no others. Colouration is variable and many individuals are much darker than the ones shown here. An upturned nose distinguishes this species from its close relative, the straight-nosed pipefish (*Nerophis ophidion*), which is less commonly seen. The male worm pipefish cares for the eggs but they are carried in a shallow groove on its belly, rather than a brood pouch, and can be very obvious as the bottom photograph shows. [Up to 15 cm long]

Aquarium photograph

Male, found on the shore, carrying eggs

Snake pipefish - *Entelurus aequoreus*

Our largest pipefish, this species is easily distinguished by its striking colouration of orange-brown with thin but clearly defined pale hoops. There is also a fetching dark stripe running back through the eye. Aside from the colour, the snake pipefish is also rather different in form to the greater pipefish and similar species. The body is much smoother and more rounded in cross section, so it does not appear armoured. There are also no pectoral fins and the tail fin is virtually non-existent (see smaller photograph), so the overall effect is certainly snake-like. Snake pipefish are often found using their tail curled round seaweed as an anchor, while they swing round sucking in their food of tiny floating animals. Rather than having a special brood pouch like that of the greater pipefish, the male snake pipefish carries the eggs in a simpler hollow on the outside of his belly. [Up to 60 cm long]

Tail used as anchor

Short-snouted seahorse - *Hippocampus hippocampus*

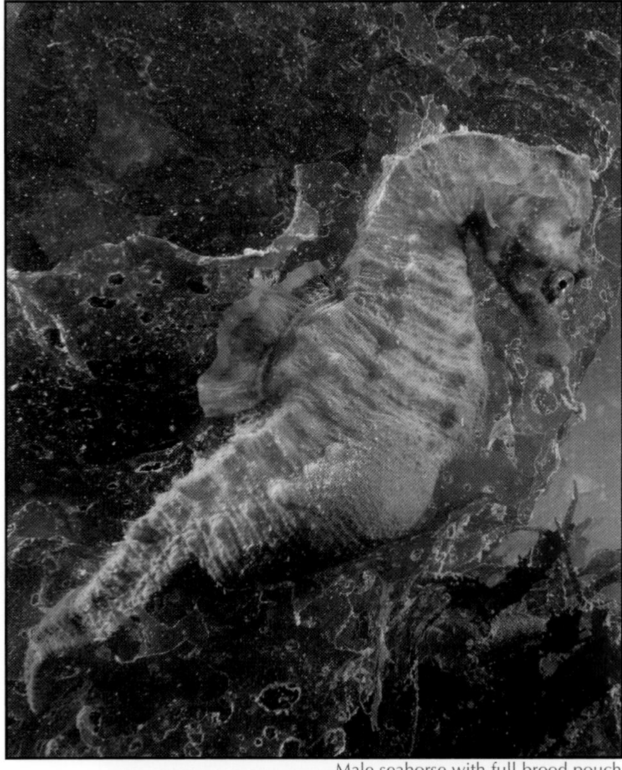
Male seahorse with full brood pouch

Sightings of seahorses are especially eagerly sought and, while this is easily explained by their special character, it shouldn't be forgotten that their much more frequently seen pipefish relatives (page 260 to 262) are wonderful too! Two species of seahorse are found in British waters; the short-snouted shown here, and the long-snouted or spiny seahorse *Hippocampus guttulatus*. Both species have the classic seahorse shape with bent over head, a clinging tail without a tail fin, and a stiff body protected by bony armour plates. The short-snouted has a slightly concave snout less than one third the length of its head; the long-snouted has a straight snout more than one third of head length. Seahorses are similar to pipefish in that the female lays her eggs into a belly pouch on the male, who then rears them and eventually releases fully-formed youngsters. Their arrangement is even more intriguing in that male and female form a monogamous pair bond which seems to be reinforced by regular greeting dances during breeding. The photographs here show a pregnant male short-snouted seahorse found near Torbay, his distended belly bearing the developing eggs is obvious. Because he stayed in the same place, I was able to see him a few times over several days and watch occasional visits by his partner and their resulting dances. The bottom photograph shows her tucked in behind him on one of the visits. More is being learnt about these fabulous animals all the time, and important work is underway to study their habits in our waters. It is vital we take a proactive approach to protecting the habitats they need to thrive. [Up to 15 cm long]

The same male seahorse with his female partner (just behind him)

Long-spined sea scorpion - *Taurulus bubalis*

The long-spined sea scorpion, sometimes simply referred to as the sea scorpion, is a common fish of shallow water and surprisingly large individuals can even be found in rockpools. The first part of its name comes from the long sturdy spine pointing backwards from each gill cover. There are also some smaller spines nearby but none are venomous, unlike those of the Mediterranean scorpion fish. A distinctive little barbel is visible at either corner of the mouth. The broad head has bony crests and there are no scales but, instead, bony plates embedded in the skin. The sea scorpion's irregular outline and mottled patterning is only part of its camouflage story, for it has an impressive ability to replicate the colour of its surroundings. It will often stay perfectly still when approached, just swivelling its eyes to assess the intruder. The varied habitats in which they live give sea scorpions plenty of opportunity to show off their colour mimicry repertoire of pinks (encrusting algae), deep reds and browns (seaweeds) and oranges (sponges). There are some examples here, although the fish in the final photograph is included because it seems to have lost the plot. The claws and legs protruding from the corners of its mouth may be the explanation. While grabbing or attempting to swallow the crab, it has moved from another position and apparently forgotten (or not had time) to blend in! The sea scorpion is a voracious predator, creeping up on crustaceans or fish and then lurching forward to take them in a single gulp. Surprisingly large prey can be taken in the wide mouth. Like a number of bottom-living fish, sea scorpions lack a swimbladder to provide buoyancy and this explains their rather ungainly movements. Breeding takes place in the spring and clusters of orange eggs are laid in rock crevices or amongst seaweeds. [Up to 20 cm long]

Long-spined sea scorpion - *Taurulus bubalis*

Orange to match sponge

Maroon to match seaweed

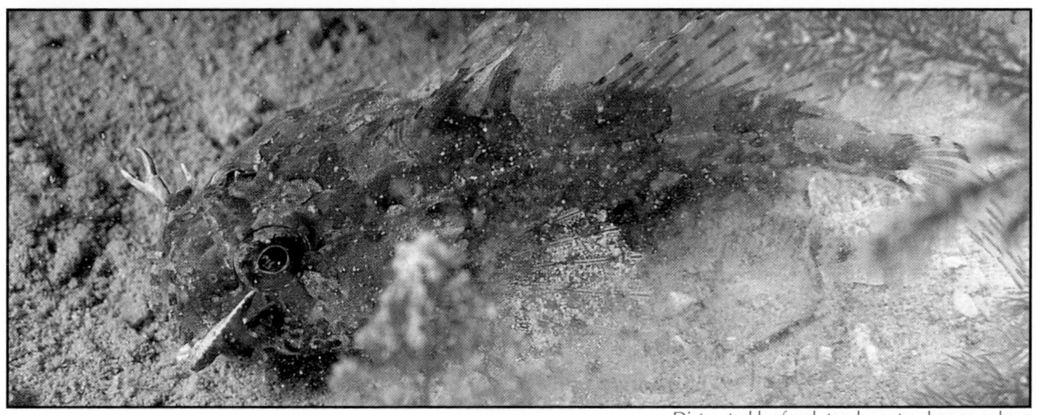

Distracted by food, too busy to change colour

Short-spined sea scorpion (or bull rout) - *Myoxocephalus scorpius*

Short-spined sea scorpion among clumps of fluted sea squirts (page 233)

This stout fish, which also goes by the fine name of "father lasher", is very similar in appearance to the long-spined sea scorpion (pages 264-265) but it can grow to a larger size. The "short-spined" name refers to the fact that the spines on its gill covers are all fairly short and there is no single long spine as in the other species. The spines are very difficult to see underwater, however, and the best distinguishing feature is that the short-spined sea scorpion has no flap of skin (barbel) at either side of the mouth. This sounds like a trivial feature but is surprisingly easy to spot underwater. Short-spined sea scorpions can be found on all types of seabed, hard or soft, and appear capable of imitating the colouration of their background without having quite the breadth of repertoire shown by their long-spined relatives. The fish in the small photograph had just grabbed a small shore crab whose legs can be seen protruding from the mouth of its nemesis. I had been watching as the sea scorpion eyed up its prey from about 40 cm away and it took the crab with a burst of startling speed. [Up to 30 cm long]

This fish had just seized a shore crab

Lumpsucker - *Cyclopterus lumpus*

Male with eggs (both photographs)

This strange looking fish has a stockily built body which is rounded and humped, and is protected by bony plates which form rows of bumps running along its length. Its head is massive and very wide. Colouration is bluish, greyish or greenish, the male taking on a reddish belly or flanks in the breeding season. There are two dorsal fins, but the foremost fin becomes overgrown by thick skin with age. The pelvic fins are fused to form a powerful sucker on the underside, just behind the chin, which is used for clinging onto rock surfaces. Lumpsuckers spend most of their time in fairly deep water but, between February and May, pairs meet up in shallow water where the female lays her eggs in a mass on a rocky ledge. She immediately returns to deeper water but the male stays to take care of the eggs for the one or two months until they hatch. It is during this time that most lumpsuckers are seen. The devoted father keeps scavenging animals away, but his chief duty may well be to keep the eggs oxygenated. He achieves this by fanning them with his fins or by pushing his head into the mass; the indentations caused by this action can sometimes be spotted (see photographs of male guarding pale yellow egg mass). Any eggs that become rotten may be eaten in order to keep the rest healthy. Egg masses are often laid in very shallow water, sometimes above low water mark, so wave surge could cause a real problem for the attendant male if he lacked the help of his sucker to keep him in position. [Females up to 60 cm long, males up to 50 cm]

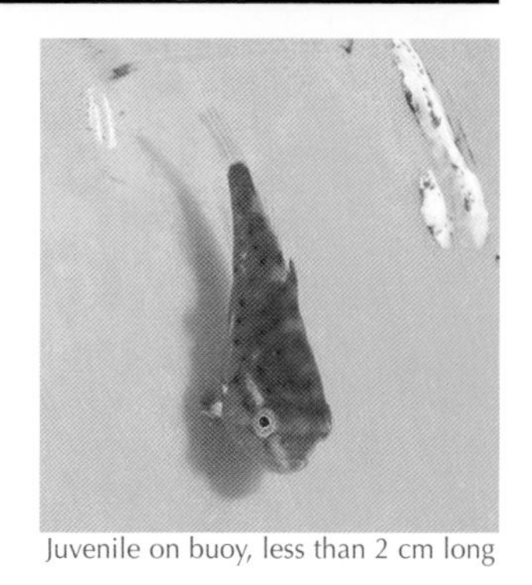

Juvenile on buoy, less than 2 cm long

Montagu's sea-snail - *Liparis montagui*

Aquarium photograph of a Montagu's sea-snail with its tail curled around

Like its much larger lumpsucker relative, this small fish has a strange appearance. Instead of appearing armoured and stocky however, it looks rather like a super-sized tadpole at first glance. A large rounded head and front-part of the body lead into a much slimmer tail. As with the lumpsucker, the pelvic fins are fused to form a sucker underneath the body, just behind the head. This is used for clinging to rocks and seaweed. The skin, which has no scales but is covered in extremely small prickles, looks slightly "baggy". Colour is usually brownish but can be very variable, depending on its surroundings. Montagu's sea-snail feeds on small crustaceans and is usually found on the shore among stones and seaweed. The closely related sea-snail (*Liparis liparis*) lives below the shore, it is very similar but the rear part of its anal fin overlaps part of the tail fin (Montagu's sea-snail has a small gap). [Montagu's sea-snail up to 6 cm long. Sea-snail up to 12 cm]

Pogge (or hooknose) - *Agonus cataphractus*

An odd but unobtrusive fish usually found on sand or mud, the pogge has a head and body completely covered with keeled, bony plates. The underside of the wide but pointed head is covered with a beard-like array of barbels that help in the finding of food. A pair of heavy curved spines on the snout gives the impression of an upturned nose. The pectoral fins are large but, behind them, the body tapers away to a very slim tail stem. Overall colour is greyish-brown with darker patches. The pogge seems content to rely on its armour and camouflage for defence, and appears unperturbed when approached. [Up to 20 cm long but usually smaller]

Grey gurnard - *Eutrigla gurnardus*

Gurnards are another group of fish whose head is armoured with bony plates, but they have an unusual profile created by a very steep forehead. Their pectoral fins are also very characteristic, with the foremost three fin rays separated from the rest of the fin to form "fingers" which probe for food and can "walk" the fish across the seabed. The grey gurnard is mainly a greyish or reddish brown with numerous small pale blotches and an off-white underside. A dark blotch is often visible on the foremost dorsal fin. Other very similar gurnard species are distinguished by brilliant blue edging and blotches on the pectoral fins (tub gurnard, *Chelidonichthys lucernus*) or an overall red colour (red gurnard, *Aspitrigla cuculus*). [Up to 45 cm long]

Streaked gurnard - *Chelidonichthys lastoviza*

This species is described as being uncommon around Britain, but my encounter with this individual (in Loch Torridon, north west Scotland) gave me a very rare opportunity to obtain a clear photograph of a gurnard. It has an even steeper forehead than the grey, red and tub gurnards (described above) and usually has a reddish back with darker blotches. The distinctive rows of spots on the pectoral fins look almost black when the fins are folded (as here) but dark blue when they are spread. It feeds on swimming crabs. [Up to 35 cm long]

Lesser sand eel - *Ammodytes tobianus*

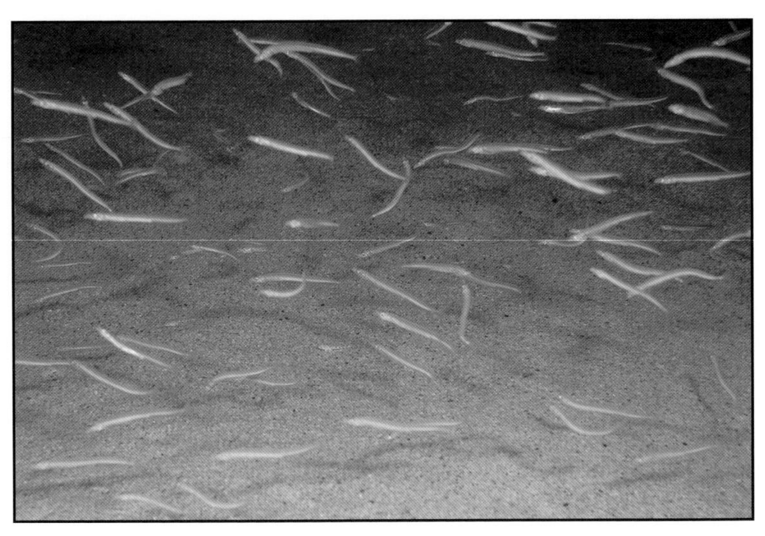

Usually seen in dense shoals in sandy areas, particularly where there are also weed-covered rocks. They are very thin, silvery fish with a jutting lower jaw and forked tail fin. Shoals can swim very quickly, however, and all that may be seen is a swirling group of silvery streaks. Some of their time is spent buried in sand and an entire shoal can disappear into the seabed in seconds. Conversely, a hand put down on the sea floor can cause a shoal to erupt into the open, a startling experience for the unwary. Shoals can be watched being followed by predators such as the greater sand eel (below), pollack (pages 252-253) and bass (opposite). They form a major food source for many fish and sea birds. There are other species of small sand eel which are very difficult to distinguish from this one. [Up to 20 cm long]

Greater sand eel - *Hyperoplus lanceolatus*

These fish are very similar in form to lesser sand eels (above) but, in addition to being larger, can appear quite different when seen underwater. They tend to swim in much smaller groups of only a few individuals and move with a far more purposeful and menacing attitude. I have encountered greater sand eels most often when they are hunting lesser sand eels. A dark spot on the snout is a distinguishing feature of the greater sand eel but they usually move too quickly for this to be noticeable other than in photographs. [Up to 35 cm long]

Bass - *Dicentrarchus labrax*

Gang of bass hunting sand smelt beneath Swanage Pier

Bass are free swimmers, normally seen in small shoals in shallow rocky areas. They are more thick-set than fish such as pollack and saithe, and have much larger heads and mouths than grey mullet, but still appear very streamlined. The flanks are bright silver with the back slightly darker. Bass have a distinctly menacing look, and give the entirely accurate impression of being fast and voracious hunters. Groups of them can be found hunting shoaling fish such as sand eels or sand smelt, and it is then that they are easiest to observe. The top photograph was taken underneath Swanage Pier, where an early morning dive allowed me to watch a shoal of sand smelt being terrorised by eager bass. When they are not hunting, bass will usually move on immediately once they are disturbed. Occasionally, a group will circle a stationary diver at a distance, with a curious individual or two swooping down to take a closer look. [Up to 80 cm long]

Horse mackerel (or scad) - *Trachurus trachurus*

Horse mackerel are open-water swimmers and, when moving fast, may be hard to tell apart from other silvery fish. When they slow down, their characteristic shape with a very slender tail stem and close-set dorsal fins can be appreciated, and the characteristic bony scales along the mid-line may be spotted. These help to distinguish them from mackerel (*Scomber scombrus*) which have sleeker, slimmer bodies and more widely spaced dorsal fins. A large tightly packed shoal of either species swirling in the sunlight is a memorable sight, but hard to capture on camera. The horse mackerel shown here were in quite a loose small shoal that were feeding around the wreck of the James Eagan Layne in Whitsand Bay, Cornwall; the stringy material in the water is plankton. [Up to 50 cm long, usually smaller]

Black sea-bream - *Spondyliosoma cantharus*

Juvenile black sea-bream can be common among eel-grass in some areas; this youngster was photographed in Dorset where I regularly see them. Characteristics of the juveniles are the dark rim around the tail fin and pretty yellow dotted lines down the flanks. Adults, which tend to be seen less often, are more of an overall silver, sometimes with vertical dark bars on their sides, while nesting males turn almost black (hence the name). Features common to all age groups are the single long dorsal fin and the small mouth. [Adults up to 40 cm long, juvenile as shown here 5 cm]

Juvenile

Red mullet - *Mullus surmuletus*

Red mullet seen on a night dive

The most distinctive features of this attractive fish are the pair of long feelers or barbels on its chin and the steep sloping forehead. The scales on its body are very obvious and the colour varies from yellow-brown to orange-red, usually with pretty stripes or patterning. Red mullet can often be seen "snuffling" around on sandy or muddy seabeds, using their barbels to feel for small buried animals on which they feed. On occasion, the lower part of their head is pushed right down into the sand (see bottom photograph). They are common in many sandy bays, often occurring in small groups. If a group is approached too closely, an individual fish raising its front dorsal fin seems to act as a signal for all the red mullet to swim off. [Up to 40 cm long but usually much smaller]

Pair of red mullet feeding (with dragonet in foreground)

Feeding enthusiastically, part of the head is submerged in the sand

Thick-lipped grey mullet - *Chelon labrosus*

Grey mullet are common but very shy fish. Divers or snorkellers can see them very close to stony beaches, sometimes in shoals, but one often gets only the most fleeting of glimpses as they disappear rapidly when approached. Groups of small individuals are found in rock pools. The grey to silvery body is torpedo-shaped with fairly distinct stripes down the flanks. If not scared off, grey mullet can be watched grazing head-down on the seabed, scraping algae off stones and rocks or sucking up mud to extract any edible matter. The intestine is very long in order to handle this austere diet. There are two other similar species, the thin-lipped and golden grey mullets. [Up to 70 cm long]

Lesser weever - *Echiichthys vipera*

The lesser weever is one of the very few venomous fish found in British waters, and it can be quite common in shallow sandy areas. Its head has an extremely deep shape and this, coupled with the upturned mouth, gives it a distinctive and very grumpy appearance. Behind the head, the body slims down and ends in rather a slender tail. Weevers spend most of their time almost completely buried in the sand with often only their eyes uncovered. When disturbed, they swim off surprisingly rapidly. On swimming off, or when disturbed while resting, the black front dorsal fin stands erect. It is this and spines on the gill covers that can inject poison if the fish is handled or accidentally trodden on. This sting is excruciatingly painful and soaking in hot water is said to break down the toxin and reduce the pain. The greater weever (*Trachinus draco*) is also venomous but is rare in shallow water. [Lesser weever up to 15 cm long, greater weever up to 40 cm]

Cuckoo wrasse - *Labrus mixtus*

Male (foreground) and female cuckoo wrasse

Cuckoo wrasse - *Labrus mixtus*

Inquisitive and territorial males (main photograph and top); unusually large group of several males and females (bottom)

Cuckoo wrasse - *Labrus mixtus*

Female

The colourful wrasse are some of the most frequently seen fish in Britain's coastal waters. Individuals of most of the five common species will be seen on many dives in rocky areas, particularly in the summer months. The cuckoo wrasse may not be the most abundant of them, but its magnificent colouration and unusual behaviour tend to make it the most notable. An inquisitive and territorial male will often swim up to a diver's mask and look them straight in the eye before following them at close quarters, even taking the odd nibble on occasion. Males have a brilliant blue head with further blue markings interspersed with orange or yellow down the flanks. All cuckoo wrasse start life as females. Females are a coral pink with a distinctive row of black and white blotches along the rear of the back. Some females turn into males later in life, depending on the proportion of the sexes in the local population, and you can occasionally see individuals at an intermediate stage between female and male colouration. Small groups consisting of a single male accompanied by one or more females are most often encountered. However, during dives off the Devon coast in the summer of 2009, I saw groups containing two or more males and several females on a number of occasions; I could not remember seeing this type of large group before. The cuckoo wrasse is usually restricted to slightly deeper water than the other wrasse species, so is seen less often when diving from the shore or snorkelling. [Males up to 35 cm long, females usually smaller]

Ballan wrasse - *Labrus bergylta*

Adult

The ballan is the largest of our wrasse, and has the appearance of being more heavily-built than the other slimmer species. Colouration is extremely variable, with many shades of brown, grey, green and red being found. There may be strong saddle-type markings, a mottled pattern or a single light stripe running the length of the body. Each of the large scales is usually relatively pale in the middle, and darker round the edge, which can give the whole fish a very spotty appearance. The young ballans, found in very shallow water and amongst seaweed in rock pools, are usually a vibrant emerald green (see smaller photograph). Like its close relative the cuckoo wrasse, the ballan always starts life as a female but some individuals become males later in life, usually after several years of breeding as a female. Unlike the cuckoo however, ballan wrasse show no obvious colour change to mark their change of sex. One can only presume it must be obvious to other ballan wrasse! Most populations have many more females than males. The ballan is abundant in most rocky areas, from very shallow water down to 20 metres depth or so. It uses its strong front teeth to prise encrusting molluscs or barnacles from the rocks and also has additional teeth in the throat for crushing its food. When accompanied by rock cooks, I have seen it take in mouthfuls of gravel (see pages 280-281). Along with many fish species, the ballan wrasse is seen much less frequently out and about in shallow water during the winter. At this time however, small groups of individuals can often be found hiding in narrow rocky crevices. [Up to 60 cm long]

Juvenile

Goldsinny - *Ctenolabrus rupestris*

This small and slim wrasse lacks the colour and pattern variations displayed by most of its relatives, and so it is easier to identify. As the name suggests, the general colouration is a pinky or reddish brown akin to gold. The belly is paler and there may be faint stripes along the flanks. Aside from the colour, a reliable distinguishing feature is the large black spot located on the top side of the tail stem. The goldsinny's overall body shape is slightly different from that of the other wrasse species, being more torpedo-shaped. It swims in the typical manner common to all the wrasse however, usually keeping its body fairly stiff and using a rowing motion of the pectoral fins for propulsion. It can be very common in rocky areas, especially around underwater cliffs or rocky drop-offs, but is less abundant in the shallowest waters of the seaweed zone than the rock cook and corkwing. Goldsinnies will often approach divers quite closely, swimming to and fro in front of them but darting for cover if they make any sudden movements. They seem very inquisitive and I have often noticed them swimming round my head, gazing intently at my air bubbles while I take photographs. When observed closely, the protruding points of their outer row of teeth can be seen. These teeth enable goldsinnies to feed on encrusting organisms, as well as small crustaceans and other bottom-living animals. They act as cleaners to larger fish (in a similar way to the rock cook, pages 280-281) and have been used to remove parasites from caged salmon in fish farms. [Up to 20 cm long]

Rock cook - *Centrolabrus exoletus*

The rock cook is sometimes the forgotten species in descriptions of the wrasse, but it is abundant in many areas (except on the east coast) and will often be seen in groups around a reef or piece of wreckage. Broadly similar in appearance to the other small wrasse, rock cooks are usually a pale reddish-brown with blue lines running along the side of the head; in the summer males have additional blue colouring down their back and flanks. The best distinguishing mark, however, is the broad dark band across the tail fin. The most notable feature of rock cook behaviour is their intriguing relationship with ballan wrasse. Rock cooks act as cleaner fish on the larger wrasse and will remove parasites from their flanks. Small groups of rock cooks can sometimes be seen escorting a single ballan, appearing to take turns in darting forward to have a quick nibble (top photograph, opposite page). Certain locations, such as the boilers on a shallow-water wreck, seem to serve as "cleaning stations" where this behaviour can regularly be observed. In some areas, particularly where patches of gravel intersperse the rocks around which the wrasse congregate, the two species also seem to feed together. A ballan hunting in the gravel for food may be surrounded by its little cohort of rock cooks looking for any scraps that are stirred up (bottom photograph, opposite page). The ballan will sometimes take a mouthful of small stones and spit them out, the rock cooks dashing in to examine material that starts to float away. See also photograph of rock cook with spider crab on page 101. [Up to 15 cm long]

Rock cook - *Centrolabrus exoletus*

Rock cook cleaning ballan wrasse

Rock cooks feeding with ballan wrasse

Corkwing wrasse - *Crenilabrus melops*

Male in front of his nest made from pink seaweed in a typical shallow area. Inset: female with blue egg-laying papilla visible

The corkwing is another small wrasse and is very abundant. It can be found in very shallow water, where it will be seen by snorkellers, and the young are common in rock pools low down on the shore. The body is quite deep when viewed in profile, but rather slim when seen head-on. Distinguishing features include stripes on the cheeks, a dark blotch in the shape of a comma behind each eye and a dark spot in the middle of the tail stem (the latter can be very difficult to see). The overall colour is variable with females generally a pale brown, and males a darker more greenish brown, with hints of blue or dark red. Breeding males have superb colouration with brilliant blue or green mixed with claret, and very prominent cheek stripes. In the spring and early summer, male corkwings can be watched bustling around the rocks collecting scraps of seaweed in their mouths, which they then ram into a crevice to create a nest. Males invite females in to lay their eggs using a courtship display, and may try their luck with more than one female. Females with eggs can be recognised by the blue egg-laying papilla (protuberance) near the anal fin (see smaller photograph). After laying, the female has no more parental involvement but the male, once he has fertilised the eggs and put more seaweed over them, guards the nest. When engaged in building or guarding activities, the normally shy corkwing becomes bold and will attempt to ward off any approach to the nest. On occasion, they seem to swim away from the nest in full view and then return surreptitiously beneath the nearby kelp, as though attempting to mislead the potential predator. [Up to 25 cm long]

Corkwing wrasse - *Crenilabrus melops*

Male placing seaweed into the nest

Male guarding nest

Connemara clingfish - *Lepadogaster candollei*

Four species of clingfish can be seen around Britain, living mainly in stony areas and within rocky crevices. Common features include their small size, a flattened head roughly triangular in shape and a single dorsal fin close to the tail. They get their name from the strong sucker formed by the merging of the pelvic fins beneath the body, just behind the head. The Connemara clingfish has a long flattened snout, that looks like a duck's bill. General colouration can vary from green to brown or red while large individuals have obvious red spots at the base of the dorsal fin (see photographs). The dorsal and tail fins are separate, unlike those of the shore clingfish (opposite top) which are joined. [Up to 8 cm long]

Shore clingfish (or Cornish sucker) - *Lepadogaster purpurea*

Photograph of clingfish uncovered on the shore. Inset: aquarium photograph

This delightful fish can be quite abundant on rocky shores. It has a very similar shape to the Connemara clingfish (opposite), with a flattened head and duck-bill snout, but can be distinguished by the two large blue spots on the top of its head just behind the eyes. The fringed tentacle in front of each eye looks impressive when in water (see inset) but lies flat out of the water where the fish will often be seen. I have never seen a shore clingfish while diving, despite spending a great deal of time in rocky shallows. When looking for animals such as clingfish under stones on the shore, it is essential to minimise any handling and return them to their exact hiding place as soon as possible, especially as they may be guarding eggs. [Up to 7 cm long]

Two-spotted clingfish - *Diplecogaster bimaculata*

The two-spotted clingfish and the very similar small-headed clingfish (*Apletodon dentatus*) share the basic clingfish features with the two *Lepadogaster* species. Their heads are not quite as flattened, however, and they have much shorter snouts without the duck-bill shape. In both species there is a short dorsal fin close to, but not joining, the tail. The male two-spotted clingfish has a distinctive purplish spot, with pale edging, on either side of its body just behind the pectoral fin (see photograph). Otherwise, it is virtually impossible to tell apart from the small-headed clingfish underwater. The general colour of both species is very variable. [Both species up to 5 cm long]

Tompot blenny - *Parablennius gattorugine*

Tompot blenny - *Parablennius gattorugine*

Tompot blenny among jewel anemones. Inset: blenny sharing crevice with a velvet swimming crab

Although fairly small, tompot blennies are among the most charismatic fish of British seas. Their distinctive clown-like faces are often seen peering out from rocky holes but they may also be encountered out in the open. Small tompots are sometimes found on the shore. The wide mouth, eyes set high up on the head and large branched tentacle above each eye all contribute to the comical appearance. There is a much smaller fringed tentacle on the nostril beneath each eye. The elongated body is taller than it is broad and the single dorsal fin almost reaches the tail. Overall colour is a reddish or olive brown with several vertical dark bands on the body. Like all blennies, tompots lack the buoyancy of a swimbladder but their wriggling swimming motion can be surprisingly rapid. A particular tompot may be spotted in the same hole on repeated dives and, if the hole is empty at first, you might be treated to the sight of the resident bustling back into its home. Tompots are very inquisitive and may even emerge from their shelter to inspect you. If they are not too busy investigating you, they can be watched interacting with other animals and each other. Because of where they live in rocky shallows, a tompot often shares a crevice with a velvet swimming crab (inset photograph above). These crabs are notoriously feisty but tompots respond very fiercely and effectively to crabs that come too close! I find interactions between tompot blennies difficult to interpret in terms of what may be territorial dispute or courtship; a few examples are shown in the photographs on page 288 and 289. These include a photograph of a pair that was seen on several occasions with eggs, even though most literature describes the male as being the sole egg guardian. Tompots eat a variety of animals from the seabed, including sea anemones which are unpalatable to most predators. Found around most of Britain, these blennies appear to be absent in much of the southern North Sea although sightings have been reported from Essex and Norfolk. [Up to 30 cm long but usually much smaller]

Tompot blenny - *Parablennius gattorugine*

This tompot blenny appeared to be making a deliberate display in front of the hole occupied by the other individual

Two blennies in a crevice, one would often leave and then return a few minutes later

Tompot blenny - *Parablennius gattorugine*

An apparent "face-off" between two tompot blennies

Another "face-off", this time near where a pair was seen with eggs

Pair of blennies, with small dark eggs visible along the bottom of the crevice

Shanny - *Lipophrys pholis*

Shanny in typically watchful pose

These shannies were unusually confident, crowding together on their rocky ledge right in front of the camera

Shanny - *Lipophrys pholis*

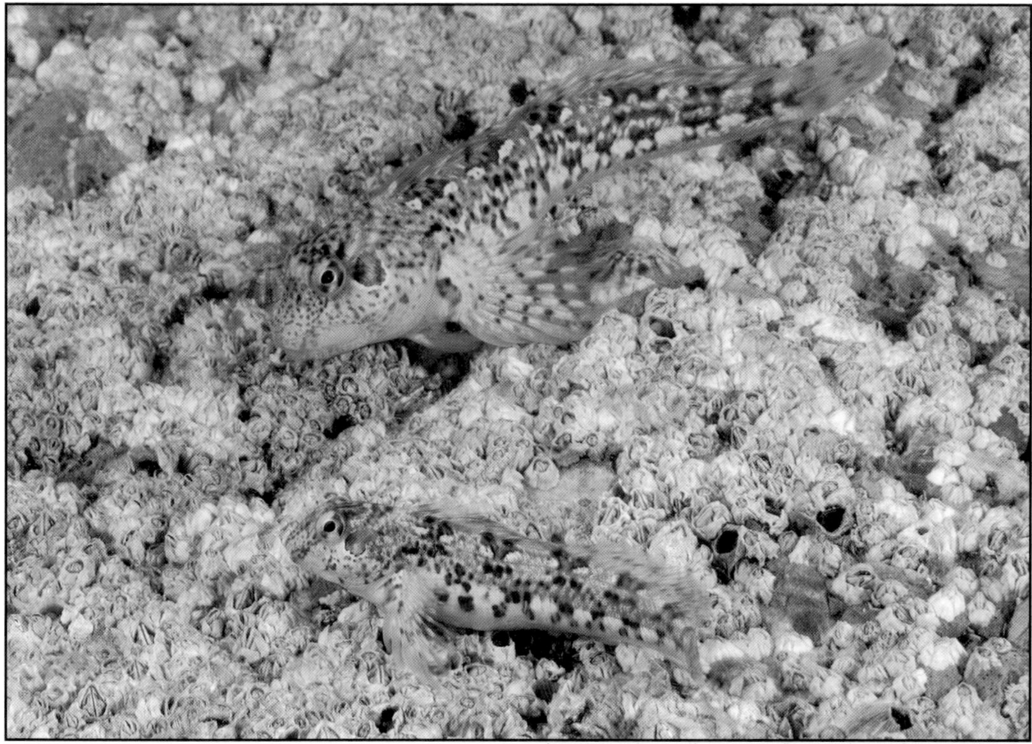

Two shannies with excellent camouflage on barnacle-covered rocks

The shanny, occasionally known as the common blenny, is a widespread and abundant fish of the shore and shallow water. It is usually seen in rock pools, by snorkellers, or by divers at the beginning or end of a beach dive. As well as hiding under stones or in holes, it can be seen in the open on barnacle covered rocks where its mottled pale brown and green colouring provides excellent camouflage (see photograph above). The shanny can also change colour to match its surroundings. It is very similar in shape to the tompot, but is usually smaller and lacks any head tentacles. It is more timid than the tompot and, when approached, it may carefully watch the viewer for just a few seconds before wriggling into a rock crack or under some seaweed. The male shanny guards the eggs, which are laid in a hole or crevice, for a month or more until they hatch. During breeding and nesting, his colour darkens to almost black while the lips are a contrasting white (see smaller photograph). Shannies are omnivorous, eating seaweed as well as small animals such as worms and shrimps. They can sometimes be watched attempting to nip off the feeding limbs of barnacles (see page 122) as these sweep out to catch food; the barnacle limbs make up a large part of the young shannies' diet. A shanny may live for as long as sixteen years. [Up to 15 cm long]

Male guarding eggs

Montagu's blenny - *Coryphoblennius galerita*

The Montagu's blenny lives in the same rock pool and rocky shore habitat as the shanny and has a similar overall appearance. It is distinguished by a fringed crest between the eyes, which is quite different from the prominent pair of head tentacles worn by the tompot blenny. Montagu's blenny also has small pale blue spots on its greeny brown head and body. Like young shannies, it eats the feeding limbs of barnacles which are abundant in the environment where it lives. This species is only found in the south and west of Britain and, even there, is seen much less frequently than the shanny or tompot. [Up to 8 cm long]

Variable (or ringneck) blenny - *Parablennius pilicornis*

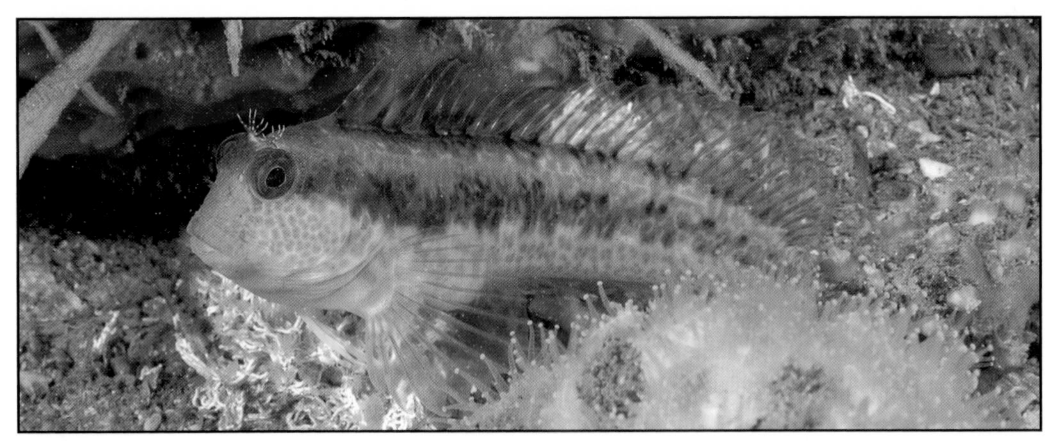

This photograph from Plymouth Sound in 2008 was thought to be only the second sighting of this species around Britain. I hesitated about including it here, but decided to do so simply as an example of the unusual sightings that can occur from time to time in our waters. Distribution in the East Atlantic is generally described as being from southern Africa to Spain.

As the "variable" name suggests, this species occurs in very different colour forms; including an overall golden yellow and a very pale overall shade with a prominent dark band running along the body. The colour form of this individual, with a dark patchy band and further yellow mottling, seems to be very characteristic of the species. [Up to 13 cm long]

Black-face blenny - *Tripterygion delaisi*

Female

Male

Black-face blennies are generally quite rare but can be a common sight in certain locations on the south coast, particularly Portland (where both these photographs were taken). I have also regularly found them in a rocky gully near Wembury in South Devon. Unlike the true blennies, which have slippery skin and a single long dorsal fin, these fish have scales and three dorsal fins. Females are a drab mottled brown (top photograph) while males are more colourful and have distinct black heads and yellow-orange bodies in the breeding season (bottom photograph). Some of the male's fins bear an attractive sky-blue edging. Black-face blennies live amongst seaweed and around rocky ledges, where they will often be found resting upside-down. The male shown here was upside-down but the photograph has been printed the "wrong" way up to make recognition easier. [Up to 9 cm long]

Yarrell's blenny - *Chirolophis ascanii*

This fish is often seen peering out from a rock crevice (top photograph) where, at first glance, it can look like a mournful tompot blenny. When found out in the open (bottom photograph), it is obviously much longer and more slender than other blennies with head decoration such as the tompot or Montagu's. If examined closely, the face alone is distinctive. A dark ring around each eye extends into a stripe down the cheek; and the head tentacles are tall with branched and very tufty ends. There are further smaller tentacles on the head and the front few spines of the dorsal fin also have tufty tips. This species is more common in the north of Britain. Like the black-face blenny, it is not classified as a true blenny. [Up to 25 cm long]

Butterfish (or gunnel) - *Pholis gunnellus*

The eel-shaped butterfish, named after its slippery skin, is distinguished by the row of white-edged black spots along the base of its dorsal fin. It is not a true blenny, but a close relative. Butterfish are often found in crevices or creeping around the base of kelp plants, sometimes lifting their heads, snake-like, to get a better view. Their eggs are laid, between December and March, in rock cavities or empty mollusc shells in shallow water. Butterfish are unusual in that the female usually guards the eggs while both parents have been reported as taking turns (in most blennies and similar fish it is just the male). They curl their bodies around the eggs and prevent them from being scattered. The butterfish is very common around all British coasts. [Up to 25 cm long]

Viviparous blenny - *Zoarces viviparus*

As the term "viviparous" in its name indicates, this fish is remarkable because it bears live young, up to 300 at a time. The eggs, instead of being shed or laid as in the majority of fish, develop within the female's body during the 3 to 4 months after mating. The young are born in the winter, about 4 cm long. It is a long slender fish, with the dorsal, tail and anal fin merged to form a single fringe. There is a distinctive notch in the top of this fringe just in front of the tail. It can be distinguished from other slender blenny-like fish because it has no obvious spots (as in the butterfish) or head tentacles (as in the Yarrell's blenny). The viviparous blenny, also known as the eelpout, is found around Scotland and right down the east coast of England. This photograph was taken in a marina in East Anglia. [Up to 50 cm long, usually 30 cm or less]

Wolf-fish - *Anarhichas lupus*

Wolf-fish are creatures of cold northern seas, and Britain is about as far south as they get. They often lurk at depths of between 60 and 300 metres but in some locations, such as on the east coast of Scotland, they can be found in as little as 10 metres of water. Here, they will usually be encountered peering out from the shelter of a cave or crevice and that fierce-looking face with prominent fang-like teeth is unmistakable. Their shape, with a large head and long tapering body with elongated dorsal fin is similar to that of an enormous blenny. Wolf-fish prey includes armoured animals such as urchins, crabs and whelks, which are crunched up with the help of strong grinding teeth on either side of the jaws and powerful cheek muscles. New teeth grow up from behind to replace the worn set every year. The same wolf-fish individuals can be seen in particular caves summer after summer, but they are assumed to move into deeper water in the winter. The female lays her eggs in the winter, producing large ball-like clumps of several thousand yellowish eggs on the seabed. After two months these hatch into larval fish which stay on the bottom while they use up their yolk reserves, but then spend some time floating with the plankton. They finally settle back on the seabed by the following autumn. [Up to 1.5 m long]

Rock goby - *Gobius paganellus*

Gobies are very abundant small fish, occasionally confused with blennies. Unlike blennies, which tend to move with a wriggle, gobies have swimbladders and swim with a more graceful darting movement. They also have two dorsal fins to the blennies' one. Gobies show similar breeding habits to blennies, with the male taking on egg-guarding duties. As its name implies, the rock goby is found on rocky ground, often in rock pools on the shore and also by divers who will see it peering out from a crevice or hole. It is fairly stout for a goby. Colour is typically mottled brown, but variable, and some fish are almost black. The front dorsal fin has a distinctive pale band along its top edge, which can be red or orange in adult males. Another distinguishing feature is the tiny branched tentacle by each nostril, although this is usually only seen when close-up photographs are examined after a dive. The smaller photograph, taken in an aquarium, shows a rock goby that had just taken a surprisingly large prawn; its tail can be seen protruding from the goby's mouth. Rock gobies are common in the south and west, but apparently very rare on the east coast. [Up to 12 cm long]

Aquarium photograph of rock goby eating a prawn

Black goby - *Gobius niger*

This species is one of the larger gobies and, like the rock goby, is stouter than most of its relatives. It lives on seabeds of sand and mud and can be common among eel-grass (see photograph). Not really black, it is found in various shades of brown with darker blotches. The front dorsal fin has some long fin rays which make it quite pointed and triangular in shape, particularly in males. There is a black mark at the front corner of both dorsal fins but these may not be obvious. The fish shown attacking then rejecting a sea slug on page 141 is a black goby. [Up to 17 cm long]

Sand goby - *Pomatoschistus minutus*

This is the small, slim sand-coloured fish so abundant in shallow sandy areas. The common goby (*Pomatoschistus microps*) is very similar. Neither species has the same obvious rows of spots on their dorsal fins as the painted goby (opposite), and they are extremely difficult to tell apart from one another. The sand goby has smaller scales down its flanks while the common goby has a triangular dark patch at the top of the pectoral fin base, which the sand goby lacks. They also prefer slightly different habitats, with the common goby living higher up on the shore and further into estuaries than the sand goby. Both species move into deeper water in the winter. [Neither sand or common gobies are usually more than 6 cm long]

Painted goby - *Pomatoschistus pictus*

Painted gobies scavenging around a feeding crab

The painted goby is slim and fawn-coloured so, at first glance, is very similar to some other gobies. It can be recognised by its dorsal fins that are decorated by one or two rows of black spots. Reddish bands on the fins above the spots are especially noticeable in breeding males. There are also dark blotches down the sides of the body, but the fin spots are much more distinctive. Painted gobies generally live in areas of coarse sand or gravel, while the closely related sand and common gobies (see opposite) live where there is fine sand or mud. The bottom photograph here shows a number of painted gobies darting expectantly around a feeding harbour crab (page 92-93), hoping for any scraps. Divers disturbing the sand will often get the same attention because they may uncover small worms or shrimps. See page 47 for an example of the gobies' unusual behaviour around a dahlia anemone. [Up to 9 cm long, usually 6 cm or less]

Two-spot goby - *Gobiusculus flavescens*

Two-spot gobies do not rest on the seabed for much of the time like other gobies, but dart continually around in the water a few cm or more above the bottom, often in patches of seaweed or eel-grass (see page 17). This habit, as much as its markings, makes the species easy to identify. General colouration is reddish brown, with a paler underside and pretty pale blue markings along the sides of the body. There is a conspicuous black spot at the base of the tail fin, and males have another dark spot on their sides just behind the pectoral fin, hence the name. In their breeding season during the spring and early summer, males develop beautifully iridescent blue lines on their dorsal fins (see bottom photograph). Females usually lay their eggs around the base of kelp plants and the males then guard them until they hatch. Two-spot gobies are usually found in small groups and these groups can sometimes be watched being tracked by predators such as young pollack (pages 252-253) or a John Dory (pages 258-259). The gobies eat floating food such as larval animals or tiny planktonic shrimps. [Up to 6 cm long]

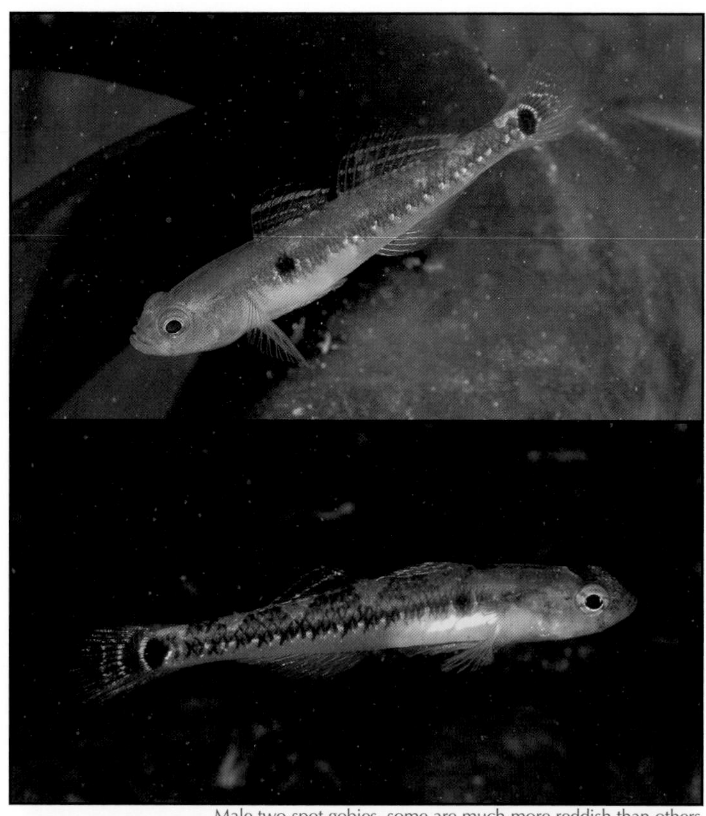

Male two-spot gobies, some are much more reddish than others

Breeding male with iridescent blue lines on its dorsal fins

Fries's goby - *Lesueurigobius friesii*

This pretty but shy species inhabits the burrows that scampi (page 118) dig in muddy seabeds, and it can be quite common where these crustaceans are abundant in Scottish west coast sea lochs. The gobies have distinctive yellow-gold spots dotted over their body and fins but they rarely grant you enough time for a close look, usually diving for cover as you approach. The fish shown in the bottom photograph, for example, darted down the nearby burrow a moment after the photograph was taken. It has been suggested that Fries's gobies may act as look-outs for their scampi hosts while benefiting from shelter and possibly extra scraps of food. [Up to 10 cm long]

Leopard-spotted goby - *Thorogobius ephippiatus*

The attractive colouration of this splendid little fish makes it easier to identify than any of the other gobies. The overall body colour is a mauve-grey or fawn, and there are numerous dark brown or brick-red blotches all over the body and head. A fetching pale blue edging to the dorsal fins is sometimes visible. When studied closely, the outlines of the large diamond-shaped scales are also quite noticeable. Far from taking pride in their appearance, leopard-spotted gobies are extremely shy fish, disappearing into rocky crevices or the cavities beneath boulders at the first sign of attention. Escape is made even quicker by their tendency, while resting on the seabed, to face towards their refuge. This attitude contrasts sharply with that of many other small fish, who seem keen to watch the world go by. Despite being so shy, leopard-spotted gobies are a frequent sight when diving in rocky areas, particularly if a little sand or silt is present around the crevices. Before diving in such areas was commonplace, it was assumed to be a very rare species because so few were caught in trawls. It is fairly widespread but very rare or absent on the east coast south of St. Abbs. [Up to 12 cm long]

Common dragonet - *Callionymus lyra*

Female or juvenile male colouration (a long dorsal fin ray may indicate it is a juvenile male)

Adult male

A slender fish seen darting away on a sandy seabed may well be a dragonet, but with this behaviour and the prominent eyes on top of the head, very small individuals could be confused with gobies. The dragonet, however, has a much broader and nearly triangular head when seen from above. The snout is also much longer and incorporates a jutting lower jaw. Females and juvenile males are usually a pale, blotchy brown, sometimes with attractive mottling or patterns (top photograph). They are also capable of blending in perfectly with a coarse sand or gravel seabed. The adult male is impressively coloured in hues of blue and yellow (bottom photograph) but is unfortunately a rare sight. Courtship apparently consists of the male performing an elaborate display, darting around in front of a female while spreading his brightly coloured fins and even pulling a strange face. Once the female is suitably impressed, they swim up towards the surface together and shed eggs and sperm into the water, their anal fins being positioned in such a way as to keep these together long enough for fertilisation to occur. Males are thought to only breed once in a lifetime. Dragonets feed on small animals in the sand such as worms and crustaceans. Individual fish react differently when encountered, some disappearing rapidly while others seem content to be approached and photographed. There are two other species of dragonet, smaller and seen less often than this one, with which identification can sometimes be confused. [Common dragonet females up to 20 cm long, males up to 30 cm]

Plaice - *Pleuronectes platessa*

A common flatfish, frequently found on sandy and muddy seabeds, or on sand patches in rocky areas. They may be seen at any depth between the shore and very deep water. Plaice are easily distinguished from relatives such as flounder and dab by the characteristic orange spots scattered over the upper side of their body. The background colour is usually grey-brown but this can change to blend in with the seabed; the underside is a pearly white. Flatfish such as plaice are of course actually lying on their sides. Newly hatched from the floating egg, tiny flatfish look like conventional fish with the body positioned vertically in the water and an eye on either side. As they develop, still in open water, the body prepares itself for the bottom-living life. By the time the young fish have settled on the seabed, one eye has moved over to join the other eye on the same side of the body, thus producing the typically twisted facial expression of all flatfish. Plaice are "right-eyed" in that the eyes both end up on what was originally the fish's right side. Female plaice lay up to half a million eggs at a time. Mating is not intimate, as eggs and sperm are simply released into the water, but the female takes care to release her buoyant eggs beneath the male so they float up through his sperm to maximise fertilisation. Plaice are usually seen only when stationary, but prey on animals such as cockles, shrimps, worms and brittle stars while cruising over the seabed. They are known to nip off the rearmost segments of lugworms (page 130) when they extend their tails from their burrows to defaecate. [Up to 90 cm long but intensive fishing has meant that such large plaice are extremely rare. Usually no more than 50 cm]

Dab - *Limanda limanda*

Flatfish are notoriously difficult to identify. A small plaice-shaped fish lacking bright orange spots, seen on a sandy seabed, will often be a dab. The dab's lateral line makes recognition easier by taking a sharp upward curve as it passes over the pectoral fin (downward curve on this photograph, taken from the fish's other side). This, however, can be very difficult to ascertain, even in a photograph. The overall colour is brown, often with dark blotches and white or yellowish spots. There may also be reddish brown spots, but these are much less striking than those of the plaice. [Up to 40 cm long, but usually less than 25 cm]

Lemon sole - *Microstomus kitt*

Despite its name, the lemon sole is shaped like flatfish such as the plaice, flounder and dab (of which it is a relative) rather than having the characteristic rounded head of the sole (see page 306). Unlike all of these other flatfish, however, it is frequently seen on rocky seabeds such as here near St Abbs. Lemon soles often have an attractive and distinctive mottling pattern, which helps them to blend in with their surroundings. They feed on small rock-dwelling animals such as barnacles (page 122) and chitons (page 144), and also on worms and the extended siphons of bivalve molluscs living in sand. [Up to 60 cm long]

Sole - *Solea solea*

The sole (also known as the common or Dover sole) has a very characteristic rounded head, with a small curved mouth that is not positioned at the end of the snout as in most other flatfish. Its colour can vary from an even grey to a blotchy brown, depending on the seabed. The pectoral fin on its upper side, which has a black patch at its tip, can be held erect while the sole buries itself in the sand; it may mimic the dorsal fin of the poisonous weever fish (page 274) and so help to repel predators. Like the plaice, dab and lemon sole, the sole's eyes are on what started out as the right-hand side of its body. [Up to 60 cm long but usually no more than 40 cm]

Solenette - *Buglossidium luteum*

As its name suggests, the solenette looks just like a miniature sole. It has a similar overall shape and the same rounded head with small curved mouth. The easiest way to distinguish a solenette from a young sole is that the solenette's fringing dorsal and anal fins appear striped. This is because every fifth or sixth fin ray is dark for most of its length. A less obvious difference is that the solenette has no black patch on its small pectoral fin. Their pale sandy colouration and small size can make solenettes very difficult to spot on the sandy seabeds where they live, especially if they are partially buried. They are, however, very appealing to watch closely. Being so thin, the solenette's body seems almost to flow over ripples in the sand as it glides along. [Up to 13 cm long]

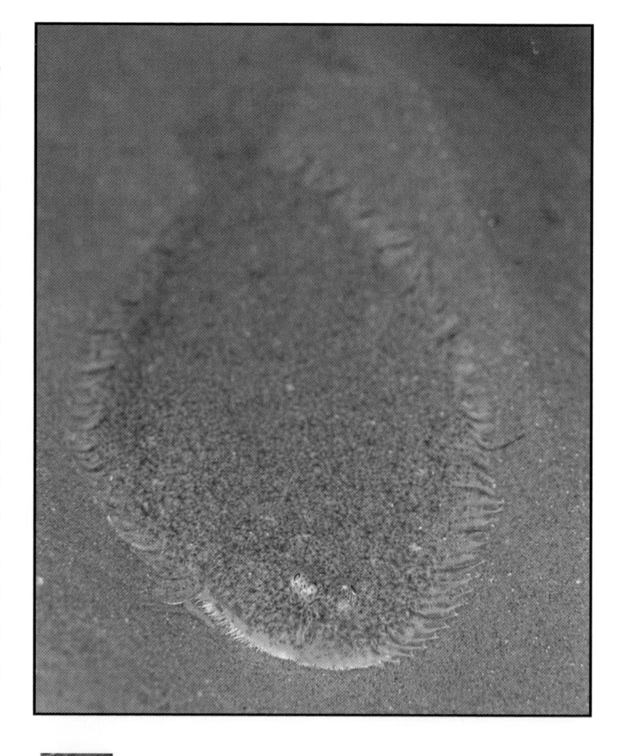

Topknot - *Zeugopterus punctatus*

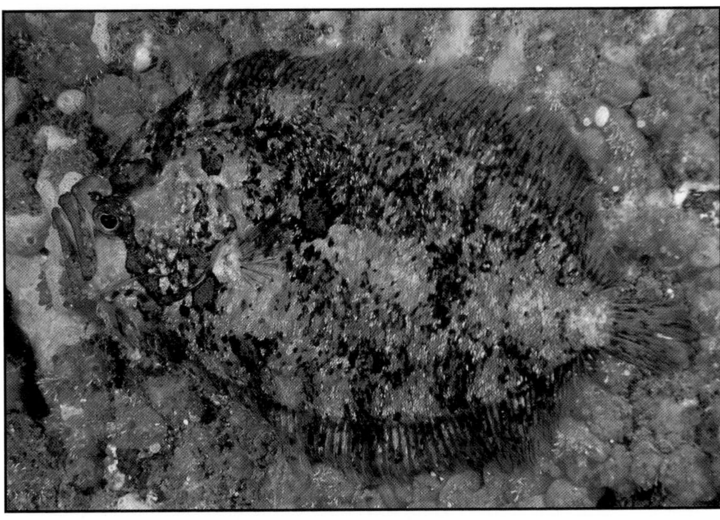

The topknot is an unusual flatfish in that it is always found on rocky ground, unlike most species which live on sand or mud. It is "left-eyed" (both eyes end up on what was the left side of young fish) and this separates it from the "right-eyed" lemon sole which is also seen on rocky seabeds. Topknots have a broad body with a fringing dorsal fin that starts next to

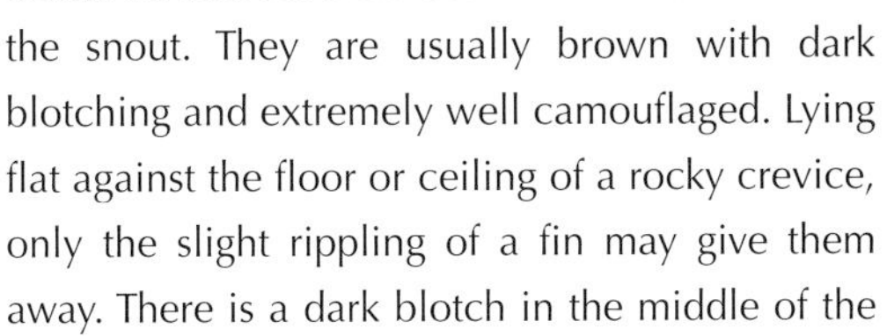

the snout. They are usually brown with dark blotching and extremely well camouflaged. Lying flat against the floor or ceiling of a rocky crevice, only the slight rippling of a fin may give them away. There is a dark blotch in the middle of the body and a distinctive broad dark stripe running outwards from each eye. These markings distinguish this species from the similar but slightly smaller Norwegian topknot and Eckstrom's topknot. [Topknot up to 25 cm long]

Triggerfish - *Balistes capriscus*

Triggerfish (sometimes called "grey triggerfish") have been seen increasingly often around the south of Britain for some years, and their distribution now extends to northern Scotland. This is thought to be one of the results of climate change and warming seas. Their oval body has a deep belly, so it is almost diamond-shaped, and is flattened from side to side. A strong spine in the front dorsal fin forms the "trigger" (see bottom photograph) and can be locked up to help wedge the fish in a crevice if it feels threatened. Triggerfish can often be quite bold but this one, part of a group in Plymouth Sound, was rather shy; divers reported them being less timid on other occasions. [Up to 40 cm long]

"Trigger" up

Glossary

I have tried to use the minimum of technical terms in this book and, where they were used, explain their meaning as I went along. Explanations of words which may be unfamiliar are given here. For names of animal groups, reference is made to the introductory text at the start of the relevant chapter. The glossary is also available on the marinephoto.co.uk web-site, where extra information will be added.

Abdomen – Tail or lower section of an animal's body, precise meaning varies between animal groups.

Acrorhagi – Also called tubercles. Small bead-like structures on some sea anemones, which contain large numbers of stinging cells.

Algae – Simple plants including seaweeds and those that grow as thin films on rocks or in the plankton.

Anus – Opening at the end of the intestine where solid waste is expelled.

Aristotle's lantern – Feeding mechanism used by sea urchins to scrape encrusting food off rocks.

Barbel – Small fleshy extension near the mouth of some fish.

Benthic – Lives on or in the seabed.

Bivalves – Type of mollusc with two halves to their shells (see page 142).

Bryozoan – Type of very small animal with tentacles, that lives in colonies (see page 190).

Byssus – Threads produced by bivalve molluscs to anchor themselves.

Calcareous – Strongly reinforced with or made of chalk (calcium carbonate).

Carapace – Part of armour that covers the back of crustaceans such as crabs.

Carnivore – Animal that eats animal flesh.

Carrion – Dead animal matter, food for scavengers.

Cartilage – Material more flexible than bone, which forms the skeletons of sharks and rays.

Cephalopods – Specialised group of molluscs containing cuttlefish, octopus and squid (see page 143).

Cerata – Small finger-like projections on the back of sea slugs.

Chromatophore – Cell containing pigment or colour. In animals like cuttlefish, these cells enable rapid colour change.

Cilia – Very small, hair-like structures that can beat rhythmically to move fluid or particles.

Cirri – Slim, claw-like structures used by feather stars to cling onto seabed.

Cnidarian – Type of animal with stinging cells. Sea anemones, corals, hydroids and jellyfish are cnidarians (see pages 36-38).

Commensal – An animal (or plant) that lives with one of a different species. There may be co-operation, but the relationship is not as close as in symbiosis (see below).

Crustacean – Type of animal with suit of armour and limbs with joints. Crabs, lobsters, prawns, shrimps and barnacles are crustaceans (see pages 84-87).

Defaecate – Release faeces, solid waste material from digesting food.

Dehydration – Dry out. Animals living on the shore are vulnerable to this, and it can kill them.

Echinoderm – Type of animal with bony plates or spines in their skin and usually no front or back end. Starfish, brittle stars, sea urchins, sea cucumbers and feather stars are all echinoderms (see pages 194-196).

Embryo – Animal in early stage of development from an egg.

Enzyme – A substance that speeds up chemical reactions (such as digestion) without being used up itself.

Fertilisation – The joining together of a female cell (egg) and a male cell (sperm) to start the development of a new individual.

Fin ray – Support for fin of a fish.

Flagellae – Similar hair-like structures to cilia (see above) but are longer and can be more complex.

Food chain, food web – Terms used to describe plants and animals in a particular environment, with each one eating the one next to it in the chain. Food web is usually a better description because the system is more complex than a single chain.

Herbivore – Animal that eats plants (including seaweeds).

Hermaphrodite – Individual that can act as both male and female.

Hydroid – Type of cnidarian (see above and page 38).

Invertebrate – Animal without a backbone. All of those described in this book apart from the fish (and the seal) are invertebrates.

Juvenile – Young animal that has the adult form (unlike larvae) but is before breeding age.

Larvae – Life stage of many animals, between the egg and adult. Larvae are distinct from adults and may look dramatically different. They can usually fend for themselves but not breed.

Madreporite – Opening into the hydraulic (water circulation) system of animals such as starfish.

Mantle tissue – Outer part of mollusc body that produces the shell.

Medusa – Free-swimming stage in the life-cycle of hydroids and jellyfish (see page 36 and polyp below). The familiar adult jellyfish are medusae.

Mollusc – Type of animal with a soft body that is usually (but not always) protected by a hard shell. Limpets, whelks, winkles, sea slugs, mussels, oysters, scallops, cuttlefish, octopus and squid are all molluscs (see pages 140-143).

Monogamous – Only breeds with one mate (partner).

Moulting – By crustaceans for example; shedding armour suit to allow growth.

Mucus – Slimy fluid produced by animals to ease movement, trap particles etc.

Nudibranch – Type of sea slug with no shell (see page 141).

Omnivore – Animal that eats both plants and animal flesh.

Oral disc – The top of a sea anemone's column (body), it has the mouth in its centre.

Parasite – Animal (or plant) that lives in or on another (the host) from which it gets food. The host may not be damaged but it suffers from providing for the parasite.

Pedicellariae – Miniature pincer-like structures on the surface of animals such as starfish and sea urchins.

Phylum – Broadest sub-division of the animal kingdom. Most of the chapters in this book correspond with a particular animal phylum.

Plankton – Plants and animals, mainly very small, that drift in the sea.

Polyp – The sedentary stage (typically attached to a rock) in the life-cycle of sea anemones, corals, hydroids and jellyfish (see page 36 and medusa above).

Radula – A ribbon-like feeding mechanism used by many molluscs (see page 140).

Reproduction, asexual – Process where animals produce new individuals (without the sexual process) by budding or splitting.

Reproduction, sexual – Production of new individuals that involves the joining together of male and female cells (see fertilisation above).

Respiration – Reaction to produce energy and support life that typically involves taking up oxygen and releasing carbon dioxide (breathing is a type of respiration).

Rhinophore – Sensory head tentacles of sea slugs.

Salinity – Concentration (strength) of the salts in seawater.

Seabed – Ground under the sea.

Sea squirt – Type of animal with adult form that typically has a sac-like body with two siphons (tubes) for taking in and expelling water (see page 229).

Siphon – Tube by which animals (such as molluscs and sea squirts) take water in and out of the body.

Species – The smallest unit used when classifying animals and plants. A species can usually be regarded as a group of individuals that can breed among themselves.

Spicules – Small, hard spiky structures that provide sponges with support and also help make them less likely to be eaten.

Sponge – Type of very simple static animal that pumps water through canals in its body to filter out food (see page 25).

Symbiosis – Close co-operation between individuals of different species, with both benefiting.

Test – Hard skeleton of a sea urchin or the covering of a sea squirt (this is also called a tunic).

Thorax – Middle section of an animal's body, precise meaning varies between animal groups.

Tube feet – Small extendable leg-like tubes used by starfish and their relatives for movement and feeding (see page 195).

Tunic – Leathery or jelly-like outer coat of sea squirts.

Vertebrate – Animal with a backbone and a skull that holds a well-developed brain. Fish, mammals, birds, amphibians and reptiles are vertebrates.

Viviparous – Giving birth to live young; embryos develop within the female's body.

Voracious – Description for predators that appear to have a large appetite.

Worm – A term used to describe long and slim animals from several different groups. See pages 124-125 for a description of the groups.

Zooids – Individual animals that make up a bryozoan colony (see page 190). The term is also used for some sea squirts.

Zooplankton – Animals, mainly very small, that drift in the sea. Some live there full-time, while others are the young stages of animals such as crabs, worms, sea snails, starfish, fish etc.

References for more information

Many of these publications were consulted during the writing of this book.

Ackers, R. G., Moss, D., Picton, B. E., Stone, S. M. K. & Morrow, C. M. (1997). Sponges of the British Isles (Sponge V) – A colour guide and working document, 1992 Edition, revised and extended 2007. Marine Conservation Society (available from www.habitas.org.uk/marinelife/sponge_guide).

Ager, O. E. D. (2001). *Funiculina quadrangularis*. The tall sea pen. Marine Life Information Network (www.marlin.ac.uk).

Alexander, R. McN. (1990). Animals. Cambridge University Press.

Angel, H. (1975). Seashore Life on Sandy Beaches. Jarrold.

Bishop, G. (2003). The Ecology of the Rocky Shores of Sherkin Island. A Twenty-Year Perspective. Sherkin Island Marine Station.

Boycott, B. B. (1958). The Cuttlefish - *Sepia*. In: New Biology, volume 25. Penguin.

Buchsbaum, R., Buchsbaum, M., Pearse, J. & Pearse, V. (1987). Animals Without Backbones, third edition. The University of Chicago Press.

Bunker, F. StP. D., Maggs, C. A., Brodie, J. A. & Bunker, A. R. (2010). Seasearch Guide to Seaweeds of Britain and Ireland. Marine Conservation Society.

Burton, M. & Burton, R. (1975). Encyclopedia of Fish. Octopus Books.

Challinor, H., Murphy Wickens, S., Clark, J. & Murphy, A. (1999). A Beginner's Guide to Ireland's Seashore. Sherkin Island Marine Station.

Clarke, B. (2002). Good Fish Guide. Marine Conservation Society.

Daly, S. (1998). Marine Life of the Channel Islands. Kingdom Books.

Debelius, H. (1997). Mediterranean and Atlantic Fish Guide. IKAN.

Dipper, F. (2001). British Sea Fishes. Underwater World Publications.

Erwin, D. & Picton, B. (1987). Guide to Inshore Marine Life. Immel Publishing.

Fish, J. D. & Fish, S. (1996). A Student's Guide to the Seashore, second edition. Cambridge University Press.

Gasca, R. & Haddock, S. H. D. (2004). Associations between gelatinous zooplankton and hyperiid amphipods (Crustacea: Peracarida) in the Gulf of California. Hydrobiologia, 530/531, 529-535.

Gibson, R., Hexstall, B. & Rogers, A. (2001). Photographic Guide to the Sea & Shore Life of Britain & North-west Europe. Oxford University Press.

Hanlon, R. T. & Messenger, J. B. (1988). Adaptive colouration in young cuttlefish: the morphology and development of body patterns and their relation to behaviour. Philosophical Transactions of the Royal Society. B, 320, 437-487.

Hanlon, R. T. & Messenger, J. B. (1996). Cephalopod Behaviour. Cambridge University Press.

Hawkins, S. J. & Jones, H. D. (1992). Marine Field Course Guide 1. Rocky Shores. Immel Publishing.

Hayward, P. J. (1994). Animals of sandy shores. The Richmond Publishing Co. Ltd.

Hayward, P., Nelson-Smith, T. & Shields, C. (1996). Collins Pocket Guide. Sea Shore of Britain and Europe. Harper Collins Publishers.

Hayward, P. J. & Ryland, J. S. (1990). The Marine Fauna of the British Isles and North-West Europe. Oxford University Press.

Howson, C. M. & Picton, B. E. (eds) (1997). The species directory of the marine fauna and flora of the British Isles and surrounding seas. Ulster Museum and The Marine Conservation Society.

Ingle, R. W. (1980). British Crabs. Oxford University Press.

Ingle, R. W. (1992). Larval stages of Northeastern Atlantic crabs. Chapman and Hall.

Irving, R. (1998). Sussex Marine Life, an Identification Guide for Divers. East Sussex County Council.

Jackson, A. (2000). *Protanthea simplex*. Sealoch anemone. Marine Life Information Network (www.marlin.ac.uk).

Kay, P. & Dipper, F. (2009). A Field Guide to the Marine Fishes of Wales and Adjacent Waters. Marine Wildlife.

Kershaw, D. R. (1988). Animal Diversity. Chapman and Hall.

Little, C. & Kitching, J. A. (1996). The Biology of Rocky Shores. Oxford University Press.

Lythgoe, J. & Lythgoe, G. (1971). Fishes of the Sea. Blandford Press.

Maitland, P. S. & Herdson, D. (editor Coates, S.) (2009). Key to the Marine and Freshwater Fishes of Britain and Ireland. A guide to the identification of more than 370 species. Environment Agency.

Manuel, R. L. (1988). British Anthozoa. The Linnean Society, E. J. Brill/Dr W. Backhuys.

Marshall, N. B. & Marshall, O. (1971). Ocean Life. Blandford Press.

Miller, S. A. & Harley, J. P. (1996). Zoology, third edition. Wm. C. Brown Publishers.

Moen, F. E. & Svensen, E. (2004). Marine fish & invertebrates of Northern Europe. AquaPress, KOM.

Muus, B. J. & Dahlstrom, P. (1974). Collins Guide to The Sea Fishes of Britain and North-Western Europe. Collins.

Oakley, J. (2010). Seashore Safaris. Graffeg.

Picton, B. E. (1993). A Field Guide to the Shallow-water Echinoderms of the British Isles. Immel Publishing.

Picton, B. E. & Costello, M. J. (1998). The BioMAR biotope viewer: a guide to marine habitats, fauna and flora in Britain and Ireland, Environmental Sciences Unit, Trinity College, Dublin.

Picton, B. E. & Morrow, C. C. (1994). A Field Guide to the Nudibranchs of the British Isles. Immel Publishing.

Ruppert, E. E. & Barnes, R. D. (1994). Invertebrate Zoology, sixth edition. Saunders College Publishing.

Shick, J. M. (1991). A Functional Biology of Sea Anemones. Chapman and Hall.

Sims, D. W. (2005). Differences in habitat selection and reproductive strategies of male and female sharks. In: *Sexual segregation in vertebrates: ecology of the two sexes,* ed. Ruckstuhl, K. E. & Neuhaus, P., pp. 127-147. Cambridge University Press.

Sims, D. W. & Quayle, V. A. (1998). Selective foraging behaviour of basking sharks on zooplankton in a small scale front. Nature, 393, 460-464.

Sims, D. W., Southall, E. J., Richardson, A. J., Reid, P. C. & Metcalfe, J. D. (2003). Seasonal movements and behaviour of basking sharks from archival tagging: no evidence of winter hibernation. Marine Ecology Progress Series, 248, 187-196.

Speight, M. & Henderson, P. (2010). Marine Ecology, Concepts and Applications. Wiley-Blackwell.

Thompson, T. E. (1976). Biology of Opisthobranch Molluscs. The Ray Society.

Thompson, T. E. (1988). Molluscs: Benthic Opisthobranchs. The Linnean Society, E. J. Brill/Dr W. Backhuys.

Tirelli, T., Dappiano, M., Maiorana, G. & Pessani, D. (2000). Intraspecific relationships of the hermit crab *Diogenes pugilator:* predation and competition. Hydrobiologia, 439, 43-48.

Waller, G. (ed), Dando, M. & Burchett, M. (1996). Sealife, a Guide to the Marine Environment. Pica Press.

Warner, G. F. (1977). The Biology of Crabs. Paul Elek (Scientific Books).

Wilson, E. (1999). *Pachycerianthus multiplicatus*. Fireworks anemone. Marine Life Information Network (www.marlin.ac.uk).

Wood, C. (2005). Seasearch Guide to Sea Anemones and Corals of Britain and Ireland. Marine Conservation Society.

Wood, C. (2007). Seasearch Observer's Guide to Marine Life of Britain and Ireland. Marine Conservation Society.

Wood, E. (ed) (1988). Sea Life of Britain and Ireland. Immel Publishing.

Wood, L. (2008). Sea Fishes & Invertebrates of the North Sea & English Channel. New Holland.

Yonge, C. M. (1949). The Sea Shore. Collins.

Yonge, C. M. & Thompson, T. E. (1976). Living Marine Molluscs. Collins.

Young, A. & Kay, P. (1994). Marine Wildlife of Atlantic Europe. Immel Publishing.

Organisations promoting conservation and awareness of marine animals

The following organisations are all involved with these aims, and are good sources of knowledge and inspiration. It is not an exhaustive list but updated information is available on www.marinephoto.co.uk

The **Marine Conservation Society** is the UK charity that cares for our seas, shores and wildlife. MCS is the voice for everyone who loves the sea, working to secure a future for our living seas, and to save our threatened sea life before it is lost forever. www.mcsuk.org

United by their vision of Living Seas, **The Wildlife Trusts'** ambition is to lead the way towards a healthy, productive and wildlife-rich future for UK coasts and seas. They are working to ensure that significant progress is made on the key issues of Marine Protected Areas (MPAs) and wildlife, fishing and seafood, marine planning and sustainable development, and legislation and policy. www.wildlifetrusts.org

The mission of **WWF** is to stop the degradation of the planet's natural environment and to build a future in which humans live in harmony with nature, by: conserving the world's biological diversity, ensuring that the use of renewable natural resources is sustainable and reducing pollution and wasteful consumption. www.wwf.org.uk

The **Shark Trust** works to advance the worldwide conservation of sharks through science, education, influence and action. Their vision is a world where sharks thrive within a globally healthy marine ecosystem. www.sharktrust.org

The **National Trust**, through its Neptune Coastline Campaign, champions our natural coastline. The Trust provides vital protection and management of habitat and wildlife on 53,000 hectares of land along 704 miles of coast in England, Wales and Northern Ireland. www.nationaltrust.org.uk

The **Seahorse Trust** was set up to preserve and conserve the natural world, especially the marine environment, using seahorses as their flagship species. They work in partnership with many organisations and people from all over the world. www.theseahorsetrust.org

Seasearch is a project for volunteer sports divers who have an interest in what they're seeing underwater, want to learn more and want to help protect the marine environment. Their main aims are to map out the various types of seabed in the near-shore zone around the whole of Britain and Ireland and record what lives in each area; thus establishing the richest sites for marine life, the sites where there are problems and the sites which need protection. www.seasearch.org.uk

The **Marine Biological Association of the United Kingdom (MBA)** (www.mba.ac.uk) is a professional body for marine scientists with world-wide membership and over 125 years of research into marine science. Its scientific staff undertake fundamental research in marine biology, provide advice and information to support marine policy and management, and develop educational resources for the public and schools. It hosts the National Marine Biological Library (www.mba.ac.uk/nmbl)

The MBA also hosts the **Marine Life Information Network (MarLIN)** (www.marlin.ac.uk). MarLIN provides detailed, quality assured information on marine life to assist agencies and naturalists engaged in environmental assessment, planning, site designation and management.

The **Scottish Association for Marine Science** is a research organisation, committed to increasing our knowledge and stewardship of the marine environment through research, education, maintenance of research infrastructure, and knowledge transfer. www.sams.ac.uk

The **Encyclopedia of Marine Life of Britain and Ireland** is a web-based photographic guide that covers a selection of the larger animals which live round the coasts of Britain and Ireland, with information drawn from a number of sources. www.habitas.org.uk/marinelife

The purpose of the **British Marine Life Study Society** is the study and conservation of the marine fauna and flora of the shore and seas surrounding the British Isles. www.glaucus.org.uk

The **Porcupine Marine Natural History Society** is an informal society that aims to promote a wider understanding of the biology, ecology and distribution of marine organisms and to stimulate interest in marine biodiversity, especially in young people. www.pmnhs.co.uk

The **National Marine Aquarium** is a charity committed to promoting a sympathetic understanding of the sea through programmes of education, conservation and research. It operates the largest public aquarium in the UK and has a mission to drive marine conservation through engagement.

Other good places to learn about marine animals include:

- **Sea Life Centres** (Birmingham, Blackpool, Bray, Brighton, Great Yarmouth, Gweek, Hunstanton, Loch Lomond, London, Oban, Scarborough, Weymouth)
- **Blue Reef Aquarium** (Tynemouth, Portsmouth, Newquay, Hastings, Bristol)
- **The Deep** (Hull)
- **Deep Sea World** (North Queensferry)
- **Blue Planet Aquarium** (Ellesmere Port)
- **Lake District Coast Aquarium**
- **Ilfracombe Aquarium**
- **Anglesey Sea Zoo**
- **Oceanarium** (Bournemouth)
- **St Andrews Aquarium**
- **Macduff Marine Aquarium**
- **Silent World Aquarium** (Tenby)
- **Exploris Aquarium** (Portaferry)
- **National Lobster Hatchery** (Padstow)
- **Fowey Aquarium**
- **Sea Life Adventure** (Southend)
- **Guernsey Aquarium**
- **London Zoo**
- **Horniman Museum** (London)
- **Liverpool Museum Aquarium**
- **The Seashore Centre** (Paignton)
- **Living Coasts** (Torquay)

Natural England, **Scottish Natural Heritage**, the **Countryside Council for Wales** and Northern Ireland's **Council for Nature Conservation and the Countryside** are Government advisory bodies that promote wildlife conservation. They also work with the **Joint Nature Conservation Committee** (JNCC) which is the UK Government's adviser on UK and international nature conservation.

Examples of different aspects of behaviour

This section lists species that give particularly good illustrations of these types of behaviour, with the relevant page numbers. Future observations will be available on www.marinephoto.co.uk

Hunting and feeding

Velvet swimming crab eating a small shore crab

Hermit crabs scavenging

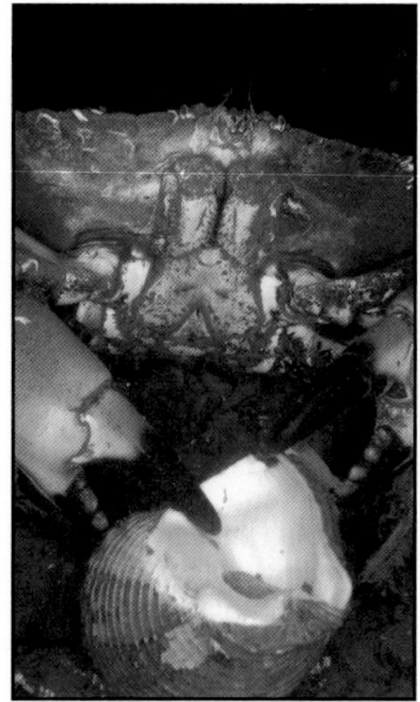
Edible crab opening a whelk

Mussels and sea squirts filter feeding

Limpet grazing algae from the rock

Cuttlefish grabbing a goby

Starfish opening a clam

Pollack hunting a shoal of sand eels

Short-spined sea scorpion eating a crab

Beadlet anemone 37
Velvet swimming crab 89
Edible crab 94
Sea toad 102
Common hermit crab 105
Long-clawed squat lobster 114
Strawberry worm 131
Limpets 146
Large necklace shell 152
Dog-whelk and netted dog-whelk 153 to 155
Sea slugs (various species) 158 to 173
Common mussel 174
Common cuttlefish 180
Lesser octopus 188
Common starfish 197
Common and purple sunstars 205 to 208
Seven armed starfish 210
Common brittle star 214
Brown sea cucumber 225
Yellow ringed sea squirt 230
Basking shark 240
Angler fish 250
Pollack 252
Sea scorpions 264 to 266
Bass 271
Red mullet 273
Rock goby 297

Defence and camouflage

Whelk with shell "door" closed

Queen scallop swimming

Shanny and dab camouflage

Coryphella browni (sea slug) gathering second-hand weapons

Heart urchin burrowing

Slender sea pen 72
Masked crab 96
Spiny spider crab 100
Long-legged spider crab 103
Common hermit crab 105
Brown shrimp 121
Coral worm 138
Elegant sea slug 141
Common whelk 156
Highland dancer 162
Coryphella browni (sea slug) 171
Common mussel 174 (154)
Gaping file shell 176
Queen scallop 179
Common cuttlefish and little cuttle 180 to 187
Common sea urchin 220
Common heart urchin 223
Cotton-spinner 224
Sprat 248
John Dory 258
Long-spined sea scorpion 264
Lesser weever 274
Shanny 290
Dab 305
Sole 306
Triggerfish 307

Co-operative or parasitic relationships between species

Cloak anemone on a hermit crab

Commensal ragworm with a hermit crab

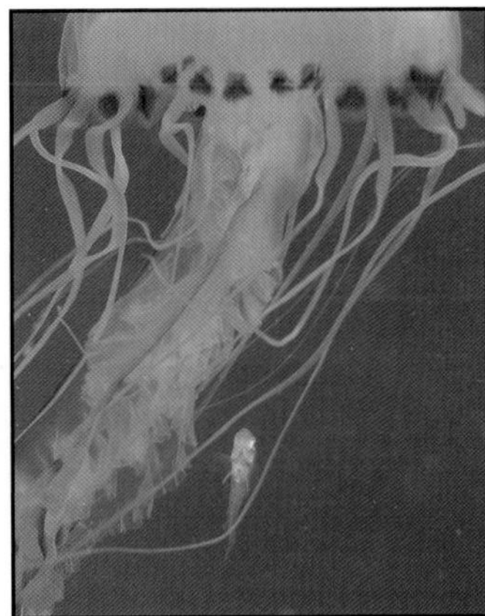
Young fish among jellyfish tentacles

Leaches on a thornback ray

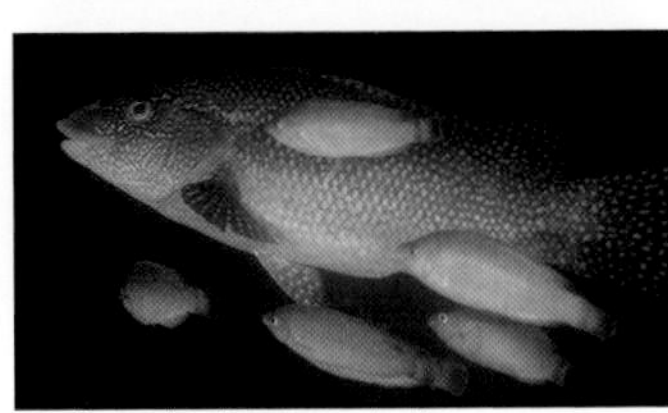
Rock cook cleaning a ballan wrasse

Breadcrumb sponge and algae 31
Snakelocks anemone and algae 42
Parasitic anemone and common hermit crab 54
Cloak anemone and anemone hermit crab 56
Hermit crab fir (a hydroid) and hermit crabs 75
Compass jellyfish and lion's mane with young fish 79, 81
Mauve stinger and amphipods 80
Leach's spider crab and snakelocks anemone 104, 42
Common prawn and lobster 119
Common prawn and conger eel 246
Sacculina carcini (a parasitic barnacle) and crabs 123
Commensal ragworm and common hermit crab 129
Leach and thornback ray 139
Queen scallop and sponge 179
Pollack and lamprey 252
Rock cook and ballan wrasse (and spider crab) 280 (100)
Fries's goby and scampi 301

Courtship and breeding, including asexual reproduction

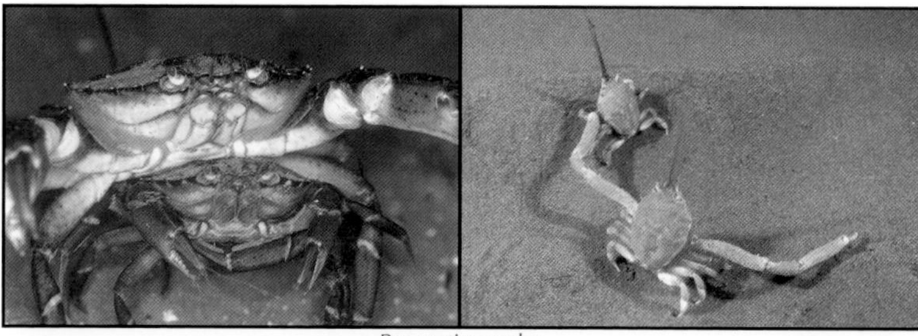
Pre-mating crabs

Asexual reproduction by sea anemones

Red whelk laying eggs

Sea hare mating chain

Starfish, sunstar and sea urchin spawning

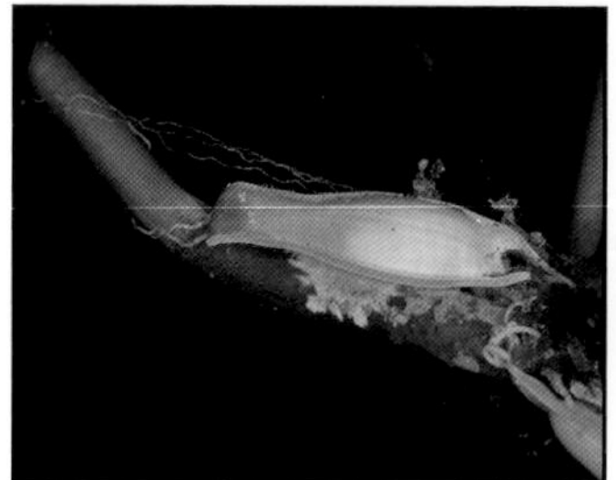
Catshark egg cases

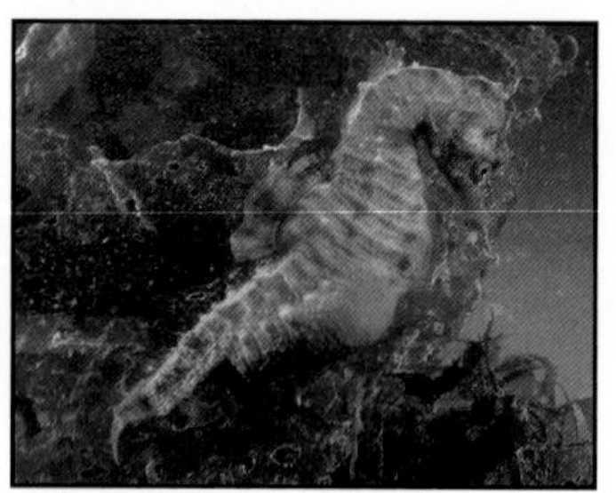
Male seahorse with full brood pouch

Corkwing wrasse building its nest

Plumose anemone 44
Sagartiogeton laceratus (a sea anemone) 52
Jewel anemone 60
Swimming, edible and masked crabs 88 to 96
Spiny spider crab 100
Flat periwinkle 150
Slipper limpet 151
Common and red whelks 156 to 157

Sea slugs (various species) 158 to 170
Common cuttlefish 180
Common and spiny starfish 197 to 203
Common sunstar 205
Sand star 212
Shore sea urchin 222
Smallspotted catshark and thornback ray 242 to 245
Worm pipefish 261
Short-snouted seahorse 263

Lumpsucker 267
Cuckoo wrasse 275
Corkwing wrasse 282
Tompot blenny and shanny 286 to 291
Viviparous blenny 295
Two-spot goby 300
Plaice 304
Typical jellyfish, crab and starfish life cycles 38, 87, 194

Index of species, English and scientific names

Acanthodoris pilosa 168 (192)
Acrocnida brachiata 219
Acteon tornatilis 158
Actinia equina 39
Actinia fragacea 41
Actinothoe sphyrodeta 49
Adamsia carciniopados 56
Aeolidiella glauca 173
Aequipecten opercularis 179
Aequorea forskalea 77
Agonus cataphractus 268
Aiptasia mutabilis 49
Akera bullata 159
Alcyonidium diaphanum 192
Alcyonium digitatum 66
Alcyonium glomeratum 68
Alcyonium hibernicum 69
Ammodytes tobianus 270
Amphilectus fucorum 32
Anarhichas lupus 296
Anemone, beadlet 39 (37)
Anemone, burrowing 59
Anemone, cloak 56
Anemone, dahlia 46
Anemone, daisy 53
Anemone, elegant 50
Anemone, fireworks 58
Anemone, gem (or wartlet) 41
Anemone, jewel 60
Anemone, parasitic 54
Anemone, plumose 44
Anemone, sea loch 58
Anemone, snakelocks 42
Anemone, strawberry 41
Anemone, trumpet 49
Anemone, white cluster 62
Anemone, white striped 49
Anemone, yellow cluster 63
Anemonia viridis 42
Angler fish 250
Anguilla anguilla 247
Antedon bifida 227
Antedon petasus 227
Aphrodita aculeata 128
Apletodon dentatus 285
Aplidium punctum 234
Aplysia punctata 160
Archidoris pseudoargus 170
Arenicola marina 130
Ascidia mentula 233
Ascidiella aspersa 233
Aslia lefevrei 225
Aspitrigla cuculus 269
Asterias rubens 197
Asterina gibbosa 209
Astropecten irregularis 212
Atherina presbyter 257
Auger shell 149
Aulactinia verrucosa 41
Aurelia aurita 78
Axinella dissimilis 30
Axinella infundibuliformis 30

Balanophyllia regia 65
Balanus perforatus 122
Balistes capriscus 307
Banded wedge shell 142
Barnacle 122
Barnacle, parasitic 123
Bass 271
Bib 255
Bispira volutacornis 134
Blenny, black-face 293
Blenny, Montagu's 292
Blenny, tompot 286
Blenny, variable (or ringneck) 292
Blenny, viviparous 295
Blenny, Yarrell's 294
Bloody Henry 204
Bolinopsis infundibulum 83
Botryllus schlosseri 237
Brittle star, black 217
Brittle star, common 214
Brittle star, crevice 218
Brittle star, sand 218
Brittle star, sand burrowing 219
Bryozoan, finger 192
Bryozoan, potato crisp 193
Bryozoan, spiral 193
Buccinum undatum 157
Buglossidium luteum 306
Bugula plumosa 193
Bull rout 266
Butterfish 295

Calliactis parasitica 54
Callionymus lyra 303
Calliostoma zizyphinum 148
Cancer pagurus 94
Carcinus maenas 88
Caryophyllia smithii 64
Catshark, smallspotted 242
Centrolabrus exoletus 280
Cereus pedunculatus 53
Cerianthus lloydii 59
Cetorhinus maximus 240
Chelidonichthys lastoviza 269
Chelidonichthys lucernus 269
Chelon labrosus 274
Chirolophis ascanii 294
Chitons 144
Chrysaora hysoscella 79
Ciliata mustela 254
Ciona intestinalis 230
Clathria atrasanguinea 31
Clathrina coriacea 35
Clavelina lepadiformis 231
Clingfish, Connemara 284
Clingfish, shore 285
Clingfish, small-headed 285
Clingfish, two-spotted 285
Cliona celata 28
Cluster anemone, white 62
Cluster anemone, yellow 63
Coalfish 254
Cod 251
Coley 254
Comb jelly 83
Commensal ragworm 129
Conger conger 246
Corella parallelogramma 232
Cornish sucker 285
Corymorpha nutans 74
Corynactis viridis 60
Coryphella browni 171
Coryphella lineata 172
Coryphoblennius galerita 292
Corystes cassivelaunus 96
Cotton-spinner 224
Cowrie, Arctic 153
Cowrie, European 153
Crab, anemone hermit 109
Crab, broad-clawed porcelain 111
Crab, common hermit 105
Crab, edible 94
Crab, furrowed 98
Crab, great spider 102
Crab, hairy (or bristly) 98
Crab, harbour 92
Crab, Leach's spider 104
Crab, long-clawed porcelain 111
Crab, long-legged spider 103
Crab, masked 96
Crab, Montagu's 98
Crab, Pennant's swimming 89

Crab, Risso's 98
Crab, scorpion spider 104
Crab, shore 88
Crab, spiny spider 100
Crab, square (or angular) 99
Crab, velvet swimming 89
Crangon crangon 121
Crassostrea gigas 177
Crawfish 117
Crenilabrus melops 282
Crepidula fornicata 151
Crossaster papposus 205
Ctenolabrus rupestris 279
Cup-coral, Devonshire 64
Cup-coral, sunset 65
Cushion star 209
Cushion star, red 209
Cuttle, little 187
Cuttlefish, common 180
Cyanea capillata 81
Cyanea lamarckii 82
Cyclopterus lumpus 267

Dab 305
Dead men's fingers 66
Dendrodoa grossularia 236
Diaphorodoris luteocincta 168
Diazona violacea 232
Dicentrarchus labrax 271
Diogenes pugilator 110
Diplecogaster bimaculata 285
Dipturus 244
Dogfish, lesser spotted 242
Dog-whelk 153
Dog-whelk, netted 155
Dragonet, common 303
Dysidea fragilis 34

Echiichthys vipera 274
Echinocardium cordatum 223
Echinus esculentus 220
Eel, conger 246
Eel, European (or common) 247
Eelpout 295
Eledone cirrhosa 188
Elysia viridis 161
Entelurus aequoreus 262
Eubranchus tricolor 173
Eunicella verrucosa 70
Eupolymnia nebulosa 131
Euspira catena 152
Eutrigla gurnardus 269

Fan worm 134
Father lasher 266
Feather star, Celtic 228
Feather star, common 227
File shell, gaping 176
Fir, hermit crab 75
Fir, kelp 75
Flabellina pedata 172
Flame shell 176
Flatworm, candy stripe 126
Flustra foliacea 192
Funiculina quadrangularis 72

Gadus morhua 251
Gaidropsarus mediterraneus 254
Galathea squamifera 113
Galathea strigosa 112
Gibbula cineraria 149
Gibbula umbilicalis 149
Gobius niger 298
Gobius paganellus 297
Gobiusculus flavescens 300
Goby, black 298
Goby, common 298
Goby, Fries's 301
Goby, leopard-spotted 302
Goby, painted 299
Goby, rock 297
Goby, sand 298
Goby, two-spot 300
Goldsinny 279
Goneplax rhomboides 99
Goniodoris nodosa 165
Gunnel 295
Gurnard, grey 269
Gurnard, red 269
Gurnard, streaked 269
Gurnard, tub 269

Halichondria panicea 31
Haliclona oculata 33
Haliclona viscosa 33
Heart urchin, common 223
Helcion pellucidum 145
Hemimycale columella 32
Henricia oculata 204
Henricia sanguinolenta 204
Hermit crab, anemone 109
Hermit crab, common 109
Hermit crab fir 75
Herring 248
Highland dancer 162
Hinia reticulata 155
Hippocampus guttulatus 263
Hippocampus hippocampus 263
Holothuria forskali 224
Homarus gammarus 116
Hooknose 268
Hornwrack 192
Horse mackerel 272
Horse mussel 176
Hyas araneus 102
Hydractinia echinata 75
Hydroid, antenna 76
Hydroid, branched antenna 76
Hydroid medusae 77
Hydroid, oaten pipe 74
Hydroid, solitary stalked 74
Hyperoplus lanceolatus 270

Inachus dorsettensis 104
Inachus phalangium 104

Janolus cristatus 165
Jellyfish, barrel 82
Jellyfish, compass 79
Jellyfish, moon (or common) 78
John Dory 258

Kelp fir 75

Labidoplax digitata 226
Labrus bergylta 278
Labrus mixtus 275
Lamprey 253
Lanice conchilega 132
Leech, marine 139
Lemon sole 305
Lepadogaster candollei 284
Lepadogaster purpurea 285
Lepidochitona cinerea 144
Leptasterias muelleri 204
Leptometra celtica 228
Leptopsammia pruvoti 65
Lesueurigobius friesii 301
Limacia clavigera 169
Limanda limanda 305
Limaria hians 176
Limpet, black-footed 146
Limpet, blue-rayed 145
Limpet, china 146
Limpet, common 146
Limpet, slipper 151
Lined polycera 169
Lineus longissimus 127
Ling 256
Liocarcinus depurator 92
Lion's mane 81
Lion's mane, blue 82
Liparis liparis 268
Liparis montagui 268
Lipophrys pholis 290

Littorina littorea 150
Littorina mariae 150
Littorina obtusata 150
Lobster, common 116
Lobster, Norway 118
Lobster, spiny 117
Lomanotus genei 164
Lophius piscatorius 250
Lugworm 130
Luidia ciliaris 210
Lumpsucker 267

Mackerel 272
Macropodia rostrata 103
Maja squinado 100
Marthasterias glacialis 202
Mauve stinger 80
Membranipora membranacea 191
Mermaid's glove 33
Metridium senile 44
Microstomus kitt 305
Modiolus modiolus 176
Molva molva 256
Moon snail 152
Morchellium argus 234
Mud runner 99
Mullet, grey 274
Mullet, red 273
Mullus surmuletus 273
Munida rugosa 114
Mussel, common (or edible) 174
Mussel, horse 176
Myoxocephalus scorpius 266
Mytilus edulis 174
Myxicola infundibulum 133

Neanthes fucata 129
Necklace shell, large 152
Necora puber 89
Nemertesia antennina 76
Nemertesia ramosa 76
Neopentadactyla mixta 225
Neoturris pileata 77
Nephrops norvegicus 118
Neptunea antiqua 157
Nerophis lumbriciformis 261
Nerophis ophidion 261
Nucella lapillus 153

Obelia geniculata 75
Octopus, common 189
Octopus, lesser (or curled) 188
Octopus vulgaris 189
Okenia elegans 166
Onchidoris bilamellata 167
Onchidoris muricata 167, 191
Ophiocomina nigra 217
Ophiopholis aculeata 218
Ophiothrix fragilis 214
Ophiura albida 219
Ophiura ophiura 218
Ostrea edulis 177
Oyster, common European
(or flat) 177
Oyster, Pacific 177

Pachycerianthus multiplicatus 58
Pachymatisma johnstonia 34
Paddleworm 128
Pagurus bernhardus 105
Pagurus forbesii 110
Pagurus prideaux 109
Palaemon serratus 119
Palinurus elephas 117
Pandalus montagui 120
Parablennius gattorugine 286
Parablennius pilicornis 292
Parazoanthus anguicomus 62
Parazoanthus axinellae 63
Patella depressa 146
Patella ulyssiponensis 146
Patella vulgata 146
Peachia cylindrica 53
Pecten maximus 178
Pelagia noctiluca 80
Pennatula phosphorea 73
Pentapora foliacea 193
Periwinkle, edible 150
Periwinkle, flat 150
Phallusia mammillata 234
Philine aperta 159
Pholis gunnellus 295
Phyllodocidae 128
Pilumnus hirtellus 98
Pipefish, deep-snouted
(or broad-snouted) 261
Pipefish, greater 260
Pipefish, snake 262
Pipefish, straight-nosed 261
Pipefish, worm 261
Pisidia longicornis 111
Plaice 304
Pleurobrachia pileus 83
Pleurobranchus
membranaceus 162
Pleuronectes platessa 304
Pogge 268
Pollachius pollachius 252
Pollachius virens 254
Pollack 252
Polycera faeroensis 169
Polycera quadrilineata 169
Polymastia boletiformis 27
Polymastia penicillus 27
Pomatoschistus microps 298
Pomatoschistus minutus 298
Pomatoschistus pictus 299
Pontobdella muricata 139
Poor cod 255
Porania pulvillus 209
Porcelain crab, broad-clawed 111
Porcelain crab, long-clawed 111
Porcellana platycheles 111
Portumnus latipes 89
Prawn, common 119
Prawn, Dublin Bay 118
Prawn, humpback
(pink or northern) 120
Prostheceraeus vittatus 126
Protanthea simplex 58
Psammechinus miliaris 222
Psolus phantapus 226

Ragworm, commensal 129
Raja clavata 244
Raspailia ramosa 29
Ray, thornback 244
Red fingers 68
Rhizostoma octopus 82
Rock cook 280 (101)
Rockling, five-bearded 254
Rockling, shore 254
Ross (or rose) coral 193

Sabella pavonina 135
Sabellaria alveolata 130
Sacculina carcini 123
Sagartia elegans 50
Sagartiogeton laceratus 52
Sagartiogeton undatus 51
Saithe 254
Salmacina dysteri 138
Sand eel, greater 270
Sand eel, lesser 270
Sand-mason 132
Sand smelt 257
Sand star 212
Scad 272
Scallop, great (or king) 178
Scallop, queen 179
Scampi 118
Scaphander lignarius 158
Scomber scombrus 272
Scyliorhinus canicula 242
Sea beard 76

Sea-bream, black 272
Sea chervil 192
Sea cucumber, brown 225
Sea cucumber, gravel 225
Sea cucumber, pink spotted 226
Sea fan, northern 71
Sea fan, pink 70
Sea fingers, pink 69
Sea gooseberry 83
Sea hare 160
Seahorse, long-snouted
(or spiny) 263
Seahorse, short-snouted 263
Sea lemon 170
Sea-mat 191
Sea mouse 128
Sea orange 26
Sea pen, phosphorescent 73
Sea pen, slender 72
Sea pen, tall 72
Sea potato 223
Sea scorpion, long-spined 264
Sea scorpion, short-spined 266
Sea slug, crystal 165
Sea slug, dead men's finger 163
Sea slug, elegant 166 (141)
Sea slug, fried egg 168
Sea slug, green 161
Sea slug, violet 172
Sea slug, white hedgehog 168 (192)
Sea snail 268
Sea snail, Montagu's 268
Sea squirt, club 234
Sea squirt, fluted 233
Sea squirt, football 232
Sea squirt, gas mantle 232
Sea squirt, gooseberry 236
Sea squirt, leathery 235
Sea squirt, light-bulb 231
Sea squirt, orange 236
Sea squirt, red 233
Sea squirt, star 237
Sea squirt, yellow ringed 230
Sea toad 102
Sea urchin, common (or edible) 220
Sea urchin, shore (or green) 222
Sepia officinalis 180
Sepiola atlantica 187
Serpula vermicularis 136
Shanny 290
Shark, basking 240
Shrimp, brown 121
Sidnyum elegans 235
Skate, common 244
Slug, white mud 159
Solaster endeca 208
Sole 306
Sole, lemon 305
Solea solea 306
Solenette 306
Spinachia spinachia 257
Spirorbis spirorbis 137
Spondyliosoma cantharus 272
Sponge, boring 28
Sponge, breadcrumb 31
Sponge, chimney 27
Sponge, chocolate finger 29
Sponge, crater 32
Sponge, elephant hide 34
Sponge, golf ball 29
Sponge, goosebump 34
Sponge, hedgehog 27
Sponge, honeycomb 32
Sponge, prawn cracker 30
Sponge, purse 35
Sponge, red encrusting 31
Sponge, shredded carrot 32
Sponge, volcano 33
Sponge, white lace 35
Sponge, yellow staghorn 30
Sprat 248
Sprattus sprattus 248
Squat lobster, long-clawed 114
Squat lobster, olive 113
Squat lobster, spiny 112
Squid, common 143
Star-coral, scarlet and gold 65
Starfish, bloody Henry 204
Starfish, common 197
Starfish, northern 204
Starfish, seven armed 210
Starfish, spiny 202
Staurophora mertensii 77
Stickleback, fifteen-spined 257
Stolonica socialis 236
Styela clava 235
Suberites ficus 26
Sunstar, common 205
Sunstar, purple 208
Swiftia pallida 71
Sycon ciliatum 35
Syngnathus acus 260
Syngnathus typhle 261

Taurulus bubalis 264
Tethya citrina 29
Thorogobius ephippiatus 302
Tonicella marmorea 144
Topknot 307
Top-shell, grey 149
Top-shell, painted 148
Top-shell, purple (or flat) 149
Tower shell 149
Trachinus draco 274
Trachurus trachurus 272
Triggerfish 307
Tripterygion delaisi 293
Trisopterus luscus 255
Trisopterus minutus 255
Tritonia hombergi 163
Trivia arctica 153
Trivia monacha 153
Tubulanus annulatus 127
Tubularia indivisa 74
Turritella communis 149

Urticina felina 46

Virgularia mirabilis 72

Weever, greater 274
Weever, lesser 274
Whelk, common 156
Whelk, red 157
Wolf-fish 296
Worm, bootlace 127
Worm, coral 138
Worm, double spiral 134
Worm, eyelash 133
Worm, fan 134
Worm, football jersey 127
Worm, honeycomb 130
Worm, organ pipe 136
Worm, paddle 128
Worm, peacock 135
Worm, red fan 136
Worm, sand-mason 132
Worm, strawberry 131
Wrasse, ballan 278
Wrasse, corkwing 282
Wrasse, cuckoo 275
Wrasse, goldsinny 279
Wrasse, rock cook 280 (101)

Xantho incisus 98
Xantho pilipes 98

Yellow edged polycera 169

Zeugopterus punctatus 307
Zeus faber 258
Zoarces viviparus 295